Studies in Systems, Decision and Control

Volume 368

The series "Studies in Systems, Decision and Control" (SSDC) covers both new developments and advances, as well as the state of the art, in the various areas of broadly perceived systems, decision making and control–quickly, up to date and with a high quality. The intent is to cover the theory, applications, and perspectives on the state of the art and future developments relevant to systems, decision making, control, complex processes and related areas, as embedded in the fields of engineering, computer science, physics, economics, social and life sciences, as well as the paradigms and methodologies behind them. The series contains monographs, textbooks, lecture notes and edited volumes in systems, decision making and control spanning the areas of Cyber-Physical Systems, Autonomous Systems, Sensor Networks, Control Systems, Energy Systems, Automotive Systems, Biological Systems, Vehicular Networking and Connected Vehicles, Aerospace Systems, Automation, Manufacturing, Smart Grids, Nonlinear Systems, Power Systems, Robotics, Social Systems, Economic Systems and other. Of particular value to both the contributors and the readership are the short publication timeframe and the world-wide distribution and exposure which enable both a wide and rapid dissemination of research output.

Indexed by SCOPUS, DBLP, WTI Frankfurt eG, zbMATH, SCImago.

All books published in the series are submitted for consideration in Web of Science.

More information about this series at http://www.springer.com/series/13304

Octavian Iordache

Polytopic Roadmaps

Springer

Octavian Iordache
Polystochastic
Montreal, QC, Canada

ISSN 2198-4182 ISSN 2198-4190 (electronic)
Studies in Systems, Decision and Control
ISBN 978-3-030-75632-1 ISBN 978-3-030-75630-7 (eBook)
https://doi.org/10.1007/978-3-030-75630-7

This Springer imprint is published by the registered company Springer Nature Switzerland AG
The registered company address is: Gewerbestrasse 11, 6330 Cham, Switzerland

Stages

In all beginnings dwells a magic force

For guarding us and helping us to live

Serenely let us move to distant places...

And life may summon us to newer races

So be it, heart: bid farewell without end

Hermann Hesse

The Glass Bead Game, New York, 1969

Preface

Complex systems exhibit hierarchical self-organization in multiple levels and scales under constraints. Higher dimensionality is significant part of the strategy for project implementation in the domain of complex industrial systems. Engineers and entrepreneurs aspire to control higher-dimensional systems. This control is accomplished by closing forward and backward procedures in cycles around the center. Consequently, both higher dimensionality and cyclicality are key strategies to face and prevail over complexity. They are clearly related to cyber-physical systems (CPS). Physical part supports the higher dimensionality, while cyber part closes the loops and ensures the cyclically controlled behavior.

Cyber-physical systems are considered to be the core and foundation of Industry 4.0.

Industry 4.0, associated with digitalizing, designates the Fourth Industrial Revolution, that follows on from technological evolutions brought about by pre-industrial human power (industry 0.0), water and steam power (industry 1.0), electric and oil power (industry 2.0), and computerized automation (industry 3.0). Blurring the lines between the material, biological, cognitive and intelligent or digital domains, the Fourth Industrial Revolution imposes the fusion of technologies. To implement industry 4.0, we need to build bridges between the various ways of understanding the reality, to develop new paradigms for engineering and scientific research that would cut across traditional disciplinary boundaries and provide models to integrate the material, biological, cognitive, and intelligence domains. Following Nicolai Hartmann's ontology, there exist four basic levels of reality: material, biological, cognitive and intelligent. All these levels or just some of them may serve for reality-inspired engineering (RIE) the source of artificiality.

The paradigm of cognitive-intelligent inspiration introduced and studied here as the basic tool for projects implementation in the domain of complex industrial systems is that of Polytopic Roadmaps.

Polytope is the general term of the sequence point, segment, polygon, polyhedron and so on. Reinhold Hoppe coined the term polytope for a convex solid in more dimensions. The label polytopic refers to all dimensions but use to be applied to four and more dimensions.

The Polytopic Roadmaps are guidelines inspired by the stages of cognitive-intelligent development, useful for learning, representing, designing, building, operating and controlling of complex engineering systems. The objective is to develop engineering operations, equipments, technologies and industries in successive stages as humans develop from sensory-motor stage of child to formal and post-formal development stages of adult. This means to extend the capacity of operations, equipments, technologies or industries to perform functions that would be associated to cognitive or intelligent tasks if performed by people. The goal can be thought of as an engineering project. But in order to imagine smart or intelligent equipments, technologies and industries, it is necessary to focus not only on the design of machines, installations and enterprises but also on the nature of the cognitive-intelligent functions to be performed.

The basic endeavor is to outline and implement the isomorphism or parallelism between the cognitive-intelligent development stages, and the smart operations, equipments, technology and industry development stages. It is the general isomorphism with respect to the higher-dimension development and cyclic control mechanisms.

In industry, technology and economy, engineers and entrepreneurs encounter a roadmap as a stepladder which they are drawn to ascend, stage by stage. It should be observed that the ascending stages are tied to descending ones in a kind of cycle or spiral. The spiral development results by superposition of linear and cyclic processes. This is described also by periodic tables. An example is represented by the successive industrialization, deindustrialization and reindustrialization stages in industrial cycles.

The 4D framework representing the industry 4.0 in previous monographs dedicated to polytopic projects is extended here to higher then four dimensions for polytopes as well as for industry stages. This means that the 4D polytope is naturally associated as reference architecture to industry 4.0, the 5D polytope to industry 5.0, the nD polytope to industry n.0 or in the other words to polytopic industry. We identify as polytopic operations, equipment, technology and industry, the n-dimensional, nD frameworks.

Polytopic Roadmaps are expected to serve as guidelines, to become powerful instruments releasing an abundance of significantly new capabilities and structures.

Polytopic Roadmaps are cognitive-intelligent inspired. These roadmaps are shared by functional organization of organisms and enterprises as informational and cognitive systems and by the operational structure of existing self-evolvable, self-sustainable, smart or intelligent systems. In order to reflect different aspects, physical, technological, industrial and social, the frameworks assign the polytopic character in the way we are looking for necessary messages into multipart objects that can be seen from different perspectives, at several conditioning levels or developmental stages successively and periodically accessed in increasing and decreasing hierarchical order.

Dedicated to the practical implementation of projects for high-dimensional complex systems, this book is divided into eight chapters. The book begins with the presentation in Chap. 1 of everything 4.0 frameworks. The higher-dimensional roadmaps beyond 4.0 are imagined. The role of doubling or twining and relating and of periodic tables is outlined. Chapter 2 contains a summary of methodology that emphasizes on iteration, dimensionality and cyclicality around the center. The cognitive-intelligent source of inspiration, the Piagetian and the neo-Piagetian development stages are briefly presented. We refer to learning, dynamic skills theory, executive control structures, model of hierarchical complexity and relational complexity. Conservative and exploratory strategies for development are introduced.

After revealing development stages as the source of inspiration, the book chapters follow the usual curriculum for chemical engineers: operations, equipments, technologies, industries and socio-economical systems. This curriculum should be renewed to take into account the advent of complexity, industry 4.0 and the Fourth Industrial Revolution.

Chapter 3 presents elementary operations of engineering interest and their relation to combinatorial species. Partitions, permutations, classifications and complexifications are the operations analyzed in this chapter. Latin squares and formal concept analysis (FCA) are involved. The role of periodic tables is underlined. Chapter 4 is dedicated to equipments able to run the presented high-dimensional operations. Multi-scale transfer equipments, cyclic operations, conveyors and assembly lines are the detailed examples. Periodic tables for equipments are proposed. Chap. 5 presents technologies. Polytopic flow-sheets, online analytical processing (OLAP), polytopic engineering designs, enterprise control systems and reality-inspired engineering (RIE) are introduced.

Concepts such as industry 4.0, industry 5.0 and polytopic industry are presented in Chap. 6. The relation between Polytopic Roadmaps, industry 4.0 toolbox and periodical tables is pointed out. Polytopic society, society 5.0, represents the subjects of Chap. 7. Society 5.0 uses the technology as basis toward a human–machine symbiosis throughout the society. Bio-Info-Nano-Cogno (BINC) technologies, Cyber-Physical Social Systems (CPSS) and ecology aspects have been discussed in this chapter. Chapter 8 highlights the connection with other systems theory projects and evaluates the integral perspective toward everything polytopic. The General Polytopic Roadmaps (GPTR), for growing complexity, are proposed as universal guidelines, as common methodologies for engineers, researchers and entrepreneurs from various domains to synthesize their thinking and capabilities into new project implementation.

Leading polytopic projects for the four levels of reality are specified. This concerns materials, biosystems, cognitive systems and intelligent systems.

The appendices refers to combinatorial species and formal concept analysis.

Introducing ground-breaking fields of most important practical and theoretical interest and roadmaps for the higher dimensional advances in complexity domains, the book will be useful to engineers, researchers, entrepreneurs and students in different branches of production, engineering, systems sciences and applied mathematics.

Montreal, Canada
January 2021

Octavian Iordache

Contents

Abbreviations

BINC	Bio-Info-Nano-Cogno
CPS	Cyber-physical system
CPSS	Cyber-Physical-Social System
DGG	Dual-graded graphs
DoE	Designs of experiments
DST	Dynamic skill theory
DT	Digital twin
FCA	Formal concept analysis
GPTR	General Polytopic Roadmaps
GST	General systems theory
MHC	Model of hierarchical complexity
MOF	Metal organic framework
OLAP	Online analytical processing
PSA	Pressure swing adsorption
PSM	Polystochastic models
PTR	Polytopic Roadmaps
RIE	Reality-inspired engineering
SoS	System of systems

List of Figures

List of Tables

Chapter 1
Beyond 4.0

1.1 Everything 4.0

The digital revolution has not been limited to industry and manufacturing but is changing all aspects of the human activity and the society development stages [5, 6, 14]. These include technology, economy, education, health, defense, society, and environment. Subsequent to the primitive society (society 0.0) and to the agricultural society (society 1.0) the invention of the coal powered steam engines turned society into industrial society (society 2.0). Wide-spread education turned it into service society (society 3.0).

The Internet, the World Wide Web, and Social Media are transforming service societies into digital societies (society 4.0).

With computers exceeding the stage of human brainpower, with intelligent service robots and cobots, and the Big Data advent, the majority of occupations in the industrial and service sectors will be modified within the next years. Most of our current institutions will fundamentally change: the way we educate Massively Open On-line Courses, (MOOC) and personalized education, the way we do research (Big Data analytics), the transportation way (self-driving cars and drones), the way of shopping (Amazon and eBay), the way of producing (3D and 4D printing), but also our health system (personalized telemedicine), and most likely politics (participation of citizens, direct democracy) and the entire economy as well (sharing economy, co-producing consumers). Financial business, which used to be the domain of banks, is increasingly replaced by algorithmic trading. How will this change the technology industry and the society is a challenging problem for the science of complexity.

Following the pre-industrial production based on human and animal power (industry 0.0), the first industrial revolution was the mechanization of production using water, wind and steam power (industry 1.0). During the nineteenth century, the industrial revolution created mass production in factories and a vast transportation infrastructure (industry 2.0). In the latter part of the twentieth century and the start

O. Iordache, *Polytopic Roadmaps*, Studies in Systems, Decision and Control 368,
https://doi.org/10.1007/978-3-030-75630-7_1

of twenty first century, the information technology revolution created personal information processing in computers (industry 3.0). The next major technological revolution will be based on an integration of information processing and material production similar to living organisms and ecosystems clearly combining these processes (industry 4.0).

Industry 4.0 is understood as a future concept for society as a whole, the so-called society 4.0 in which people, persons, more than ever, are in front position [11].

The increasing diversity of products with short delivery cycles and simultaneously decreasing numbers of personnel available can present an additional challenge for many companies. It is important that the production of the future is moving closer to people.

This will require different logistics concepts for production supply and disposal. People are not being disregarded, quite the opposite in fact. Their requirements will be taken into account to a much greater extent in corporate planning in the future.

The industry 4.0 ecological system consists of smart factories and intelligent products with a memory that control production. It is a question of allowing people to perform high quality and creative work and giving them the opportunity to achieve a work/life balance with just as much flexibility as the production systems of the future that will be controlled by people [14].

Table 1.1 outlines some basic components of the Everything 4.0 vision.

On the rows of the Table 1.1 there are different domains of interest as for instance agriculture, coordination, economy, industry, society and so on. On the columns there are the five developmental stages. The period for development is very different from centuries to few months.

Each column of the Table 1.1 indicates the critical factors in each developmental stage [12, 13]. The five successive stages are denoted: 0.0, 1.0, 2.0, 3.0 and 4.0. Usually the lowest stage 0.0 corresponds to the natural, communal or primitive level. The highest stage 4.0 represents the vision of the fourth industrial revolution, industry 4.0 and society 4.0.

Experts of the domains may specify the five stages of development.

If, for example, we think about coordination the community stage 0.0, turned into hierarchical control 1.0, into market competition 2.0, into networks negotiation 3.0, and this will evolve into awareness based followed by collective actions 4.0. If we consider as another example the economy, the traditional primitive economy 0.0, turned into agricultural economy 1.0, this into industrial economy 2.0, this into service economy 3.0, and this evolves into digital economy 4.0.

If we focus on the education, the traditions learning 0.0, turned into memorization learning 1.0, this into open access to libraries and education 2.0, this into knowledge production learning 3.0, and this will evolve into innovation production or creative learning 4.0.

If we refer to industry we may associate the pre-industrial stage industry 0.0 to manual energy, industry 1.0 to mechanical energy, industry 2.0 to electricity and petrol, industry 3.0 to electronics and automation and industry 4.0 to digitalization.

If we consider the web internet we may associate web 0.0 to texts and human direct interactions, the web 1.0 to read capability, the social web 2.0 to

Table 1.1 Everything 4.0

	Agriculture	*Capital*	*Computing*	*Consummation*
0.0	Natural	Natural	Manual	Survival
1.0	Subsistence	Human	Main frames	Traditional needs driven
2.0	Industrial	Industrial	Personal computer	Consumerism mass consume
3.0	Selective cultures	Financial	Internet, IoT	Selective Conscious
4.0	Collaborative conscious	Cultural Creation serving	Self-learning Robots	Collaborative Conscious
	Coordination	*Defense*	*Design*	*Economy*
0.0	Community	Human manual	Individual	Traditional
1.0	Hierarchy control	Weapons	Traditional artifacts	Agriculture
2.0	Market competition	Machines	Products services	Industrial
3.0	Networks negotiation	Systems Automation	Organizational transformation	Service
4.0	Awareness collective action	Unmanned self-organized robots drones	Social transformation	Digital
	Education	*Energy*	*Food*	*Health*
0.0	Tradition	Natural	Natural	Tradition
1.0	Memorization	Utilities	Cultivation	Examination dialogue
2.0	Open access libraries internet	Personal Driven	Mechanization manufacturing	Open access internet
3.0	Knowledge producing	Renewable	Processing genetics	Semantic personalized
4.0	Innovation producing	Smart grids self-organized	Collaborative integrative	Symbiotic conscious
	Industry	*Labor Work*	*Leadership*	*Marketing*
0.0	Manual	Self sufficiency	Community	Individual
1.0	Mechanic	Slave	authority	Product centric
2.0	Electric	Labor Commodity	Incentives	Customer Centric
3.0	Electronic automation	Labor regulated Commodity	Participative	Human centric
4.0	Internet digital informational self-organized	Social business Entrepreneur	Co-creative participation	Self-Actualization

(continued)

Table 1.1 (continued)

	Agriculture	*Capital*	*Computing*	*Consummation*
	Mobility	*Nature*	*Organization*	*Ownership*
0.0	Individual	Mother Nature	Community	Communities
1.0	Public	Resources	Entrepreneur	State
2.0	Private cars	Commodity land materials	Professional	Private
3.0	Driver-less vehicles	Regulated Commodity	Shared values	Mixed public/private
4.0	Autonomous transport systems	Eco-system Commons	Project product Facing-strategy	Shared common
	Social Stage	*Technology*	*Web Internet*	*Work Security*
0.0	Communal	Indigenous	Human direct	Individual
1.0	State centric	Tools Mechanical	Read	Protect worker
2.0	Free market Ego-centric	Machines Industry	Social: read/write	Social security
3.0	Social market Regulated	System centric Automation	Semantic read/write/execute	Social market security
4.0	Co-creative Eco-centric	Human centric Information	Symbiotic read write/execute/concurrency	Self-organized Supply market
	Constructions	*Cybernetics*	*Reliability*	*University*
0.0	Communal	Cyber-less	Practical	Traditional
1.0	Private	Homeostatic feedback	Specifications	Metaphysical
2.0	City regulated	Reflexive Self-organization	Adaptive	Research
3.0	Private/city regulated	Virtual predictive	Systems, predictive	Entrepreneurial
4.0	Smart home	Embodied self-evolving	Self-evolvable, intelligence	Self-Sustainable

read/write, the semantic web 3.0 to read/write/execute and the symbiotic web 4.0 to read/write/execute and concurrency capability [3].

If we refer to social stage, the communal stage 0.0 based on natural resources is the primitive stage. State centric systems are associated to stage 1.0. In the stage 2.0 economies move from state-centered to market. The stage 3.0 corresponds to social market economies when technology emerges as critical factor.

The stage 4.0 outlines the role of co-creative and ecology centric self-organization.

As resulting from Table 1.1 for different domains, the stages 4.0 highlight the role of self-organization, self-evolvability, cooperation, collaboration, co-evolution and connectivity. This stage is frequently identified as "Self" in Polytopic Roadmaps implementations [7–9].

1.2 Polytopic Roadmaps

1.2.1 Roadmaps for Growing Complexity

The dimension of a space or object is informally defined as the minimum number of coordinates needed to specify any point within it. Thus a line has a dimension of one because only one coordinate is needed to specify a point on it. A surface such as a plane or the surface of a cylinder or sphere has a dimension of two because two coordinates are needed to specify a point on it, for example, both a latitude and longitude are required to locate a point on the surface of a sphere. The inside of a cube, a cylinder or a sphere is three-dimensional because three coordinates are needed to locate a point within these spaces.

Polytopic Roadmaps refer to all dimensions but focus on dimension four and higher than four [4].

The supplementary dimensions may be assigned to time, temperature, pressure, color, property, rank, stage or any supplementary feature, in addition to the length, width and height that conventionally describes objects. The 4D framework was used as reference architecture for industry 4.0 [9, 10].

The issues raised by Polytopic Roadmaps concern the basis of material or device structures, the hardware and software, and also the framework of scientific and engineering methods.

A method to increase dimension, step by step, is the doubling and relating illustrated in Fig. 1.1. Doubling or twining endorses dimensionality while relating endorses cyclicality.

Figure 1.1 shows a roadmap from 0 to 4D.

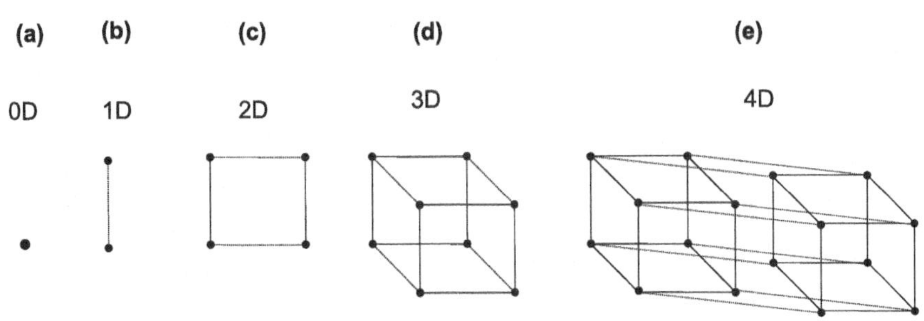

Fig. 1.1 Polytopic roadmaps: doubling and relating

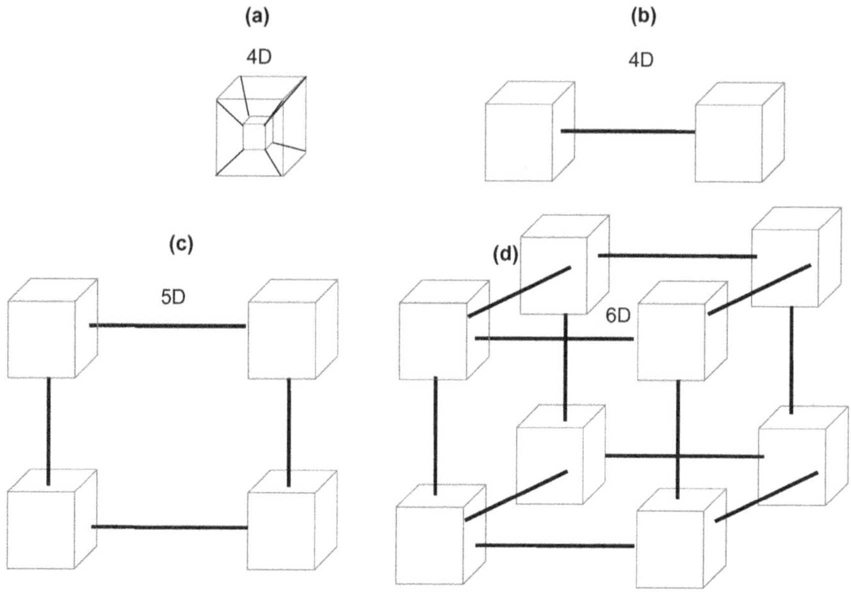

Fig. 1.2 Polytopic roadmaps from 4 to 6D

By doubling and relating the 0D dot, shown in Fig. 1.1a became a 1D segment (Fig. 1.1b). This segment became the 2D square (Fig. 1.1c), this square became the 3D cube (Fig. 1.1d) and this cube became the 4D cube (Fig. 1.1e).

Examples of Polytopic Roadmaps for growing complexity to four and higher than four dimensionality are shown in Fig. 1.2. Roadmaps from 4 to 6Dpolytope are considered in this case.

This highlights the trend towards higher dimensionality by n-cubes representations.

The continuous grows of complexity outlined the transition from evolvability frameworks (3D) to self-evolvability frameworks (4D), that is to systems that self-configure, self-optimize, self-control, self-heal, self-direct and so on, based on a set of higher dimensionality intrinsic capabilities and meeting of the user-specified variable objectives [8].

The polytope shown in Fig. 1.2a and Fig. 1.2b illustrates the frameworks of potentially self-evolvable systems. The representations are known as hypercubes or 4D cubes and consist of pairs of 3D cubes. The 4D cube is obtained by joining all corners of the first 3D cube with the corresponding corners of the second 3D cube. The doubling of a 3D cube followed by relating of corners is the source of 4D cube. Figure 1.2a shows the first cube introduced inside the second cube. Figure 1.2b shows the two 3D cubes separated.

The eight lines connecting the two 3D cubes are represented in Fig. 1.2b by a thick hyper-line. In this way the resulting 4D cube allows a simplified representation. All dimensions higher than three may be represented using such thick hyper-lines in

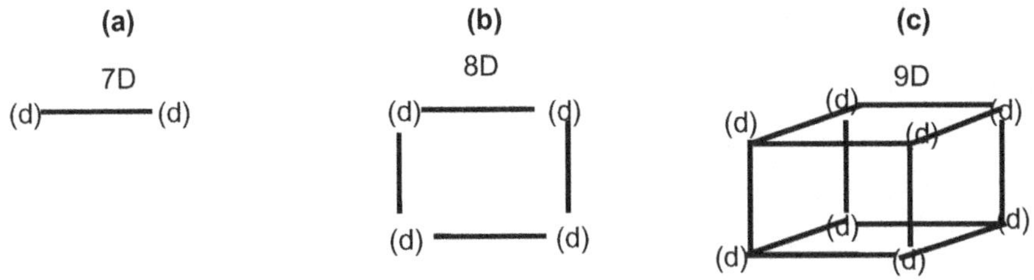

Fig. 1.3 Polytopic roadmaps from 7D to 9D

order to visually simplify the resulting structures [1, 16]. Using the thick hyper-line the 4D cube is shown in Fig. 1.2b as a line of cubes. The dimension is $1 + 3 = 4$. In this sum 1 corresponds to line while 3 correspond to cubes.

The first cube shown in Fig. 1.2b would coordinate the four stage evolvable frames shown on different faces of the second cube. Swinging, mediated by the first cube between the two complementary faces of the second cube is necessary for reuniting different trends and for self-evolvability.

The doubling and relating procedure allowing to obtain (n + 1) D frames from nD frames may continues for higher dimensionality, but the resulting polytope is more difficult to draw and to interpret. There is no fixed limit for growing complexity. So, higher dimensionality polytopes as for instance 5D and 6D cubes and others polytopes may be considered as candidate for frameworks. Figure 1.2c shows a 5D cube. Figure 1.2d shows a 6D cube. The internal cubes again should have the corresponding corners connected as for the 4D cube. Using the thick hyper-lines, the 5D cube is shown as square of cubes. The dimension is $2 + 3 = 5$ where 2 correspond to square or rectangle and 3 to cubes. The 6D cube is shown as cube of cubes of dimension $3 + 3 = 6$.

Figure 1.3 shows Polytopic Roadmaps for growing complexity from 7D to 9D.

Figure 1.3a shows the 7D case and is obtained by doubling of (d) from Fig. 1.2. That (d) is a 6D frame. The dimension for Fig. 1.3a is $1 + 6 = 7$ where 1 corresponds to the line and 6 to the 6D frames denoted by (d). Figure 1.3b shows the 8D and is obtained by doubling of (a) from Fig. 1.3. The dimension for Fig. 1.3b is $2 + 6 = 8$ where 2 correspond to the rectangle or square and 6 to the 6D frames. Figure 1.3c shows the 9D case and is obtained by doubling of (b) from Fig. 1.3. That (b) is an 8D frame.

The dimension for Fig. 1.3c is $3 + 6 = 3 + (3 + 3) = 9$. It is a cube of cube of cubes.

1.2.2 *Different Presentations*

Figure 1.4 shows several presentations of the 4D cube from Fig. 1.2 b.

The 4D cube results by doubling or twining of a 3D cube.

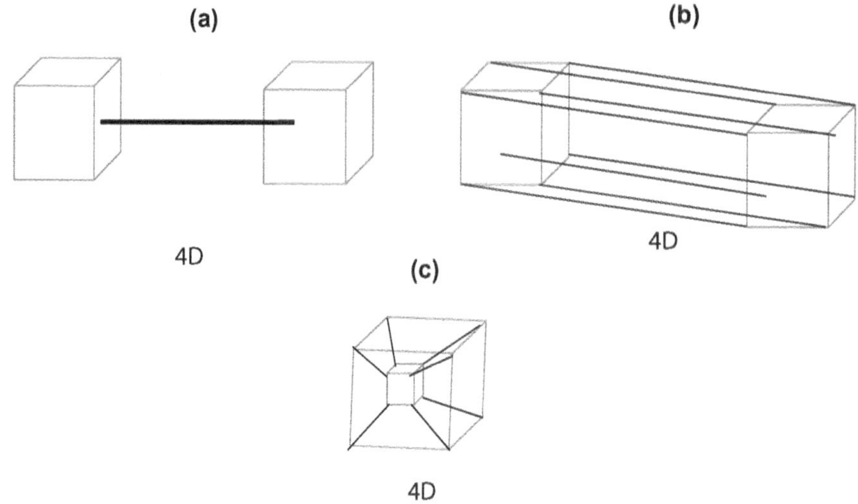

Fig. 1.4 The 4D cube

Figure 1.4a shows two 3D cubes connected by a thick hyper-line. This thick hyper-line is a simplified representation of eight connections of the corresponding corners as shown in Fig. 1.4b.The three presentations of the 4D polytope shown in Fig. 1.4 are equivalent as dimensionality. They show two 3D cubes and eight connections.

Figure 1.4a corresponds to lower density while Fig. 1.3c corresponds to higher densities. This is significant for new materials fabrication.

Figure 1.5 shows detailed presentations of the 5D cube. This is equivalent to

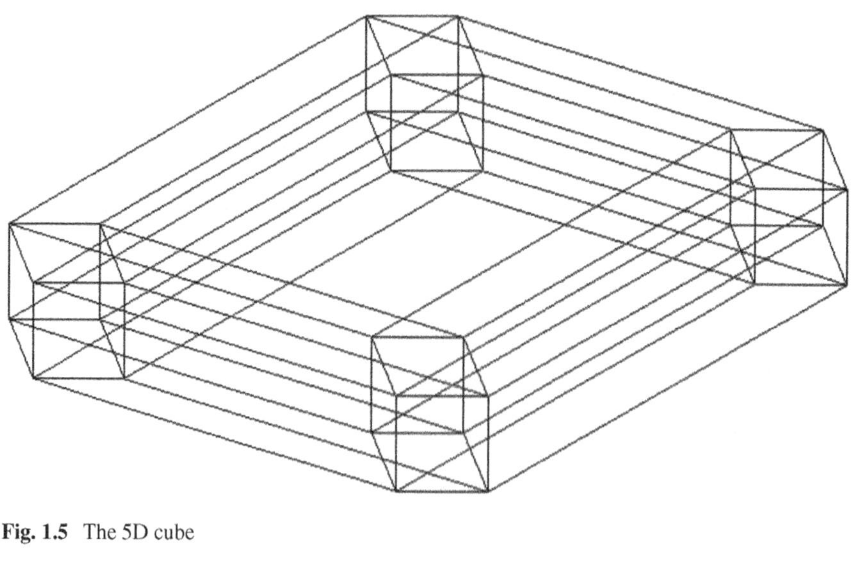

Fig. 1.5 The 5D cube

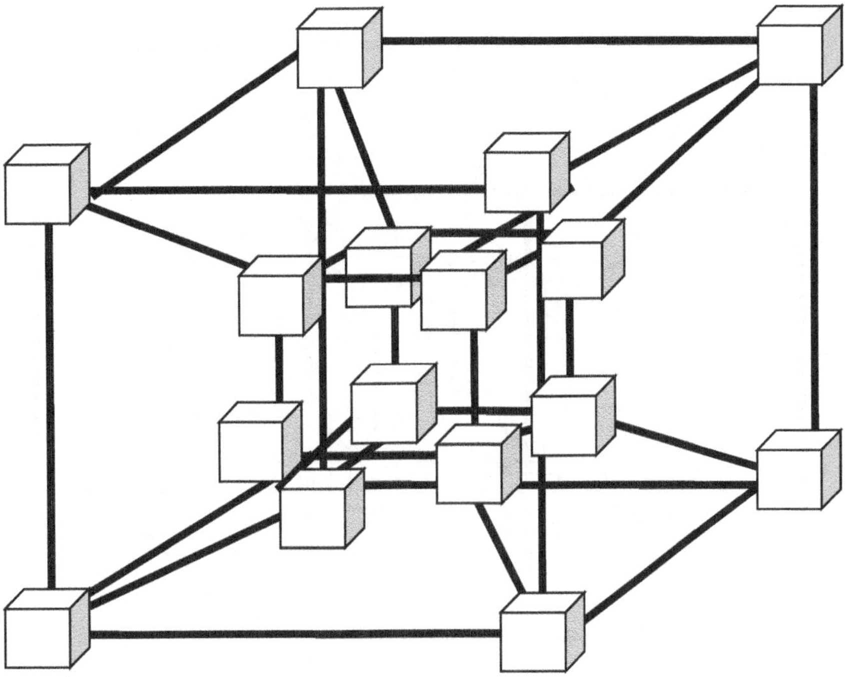

Fig. 1.6 Hypercube of cubes

Fig. 1.2c.

The 5D cube results by doubling and relating of two 4D cubes.

Frames for higher dimensionality may be of interest too.

Figure 1.6 shows a hypercube of cubes. It is a presentation of the 7D cube.

The dimension results from the hypercube (4D) having as components the eight cubes of the external frame (3D). The dimension for Fig. 1.6 is $4 + 3 = 7$. Here 4 corresponds to the hypercube shown by thick-lines while 3 to the cubes from the corners of the hypercube.

Figure 1.7 shows a hypercube of hypercubes. It is a representation of the 8D cube.

The dimension 8D for Fig. 1.7 is $4 + 4 = 8$.

Here 4 corresponds to the large hypercube shown by thick-lines while 4 corresponds to the little hypercubes from the corners of the large hypercube.

The internal little cubes allows the control of the stages. It results that the frameworks from Fig. 1.7 will be appropriate for systems with distributed control.

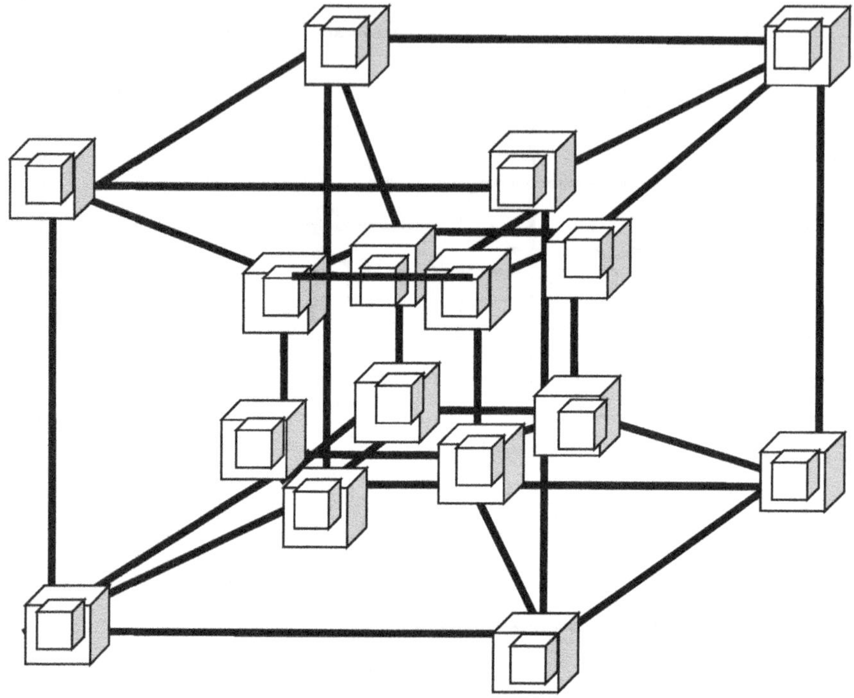

Fig. 1.7 Hypercube of hypercubes

1.2.3 Concept of Time

In classical mechanics, space and time are different categories and refer to absolute space and absolute time. The four dimensions of space time consist of events that are not absolutely defined spatially and temporally, but rather are known relative to the motion of an observer. Minkowski space first approximates the universe without gravity. The pseudo-Riemannian manifolds of general relativity describe space–time with matter and gravity. Ten dimensions are used to describe string theory, eleven dimensions can describe super gravity and the state-space of quantum mechanics is an infinite-dimensional function space.

The concept of dimension is not restricted to physical objects. High-dimensional spaces frequently occur in engineering and sciences. They may be parameter spaces, configuration spaces or reference architectures. These are abstract spaces, independent of the physical space we live in.

For some modern research in physics, high-dimensionality, that is, the new dimensions are more related to time intuition rather then to space intuition [15].

Figure 1.8 illustrates possible interpretations of the concept of time for higher dimensions.

Observe that the basic units of different frameworks are 3D cubes.

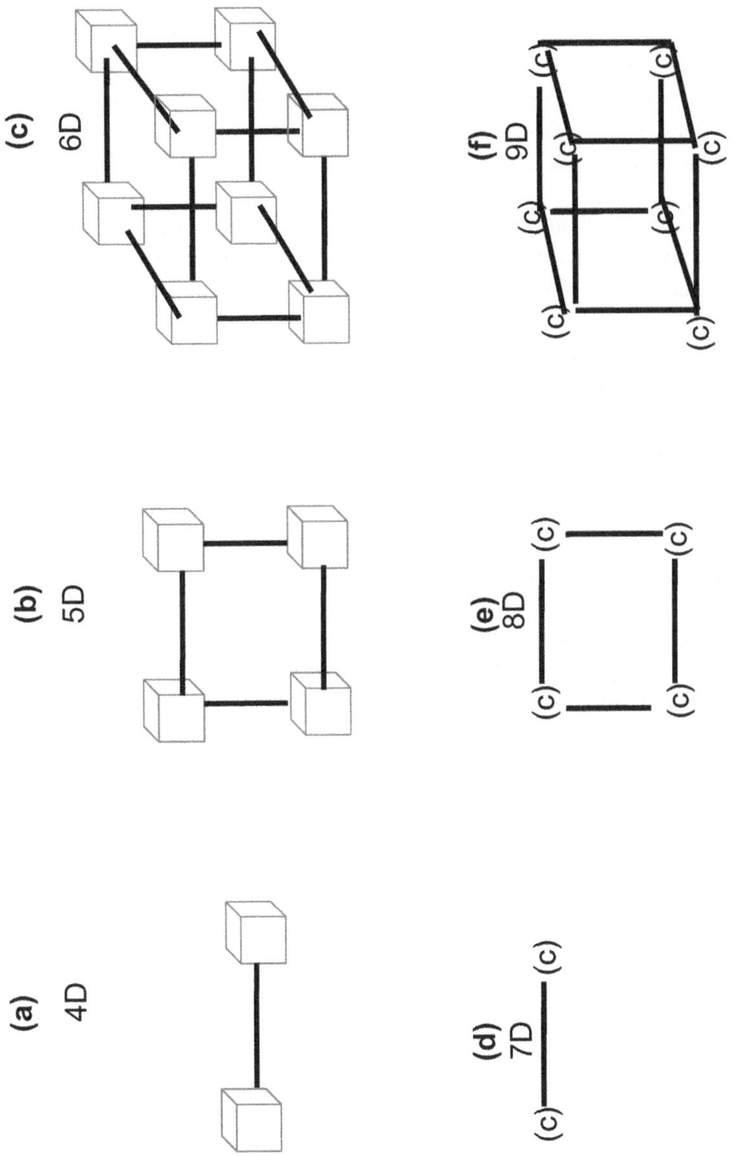

Fig. 1.8 Polytopic roadmaps for concept of time

The 4D case, as shown in Fig. 1.8a, may be associated to the linear time. For instance the 4D printing refers to printing of 3D objects that after printing varies in time [17]. It is the linear time revealed by the thick hyper-line in Fig. 1.8a, that is the line of cubes. The dimension for Fig. 1.8a is $1 + 3 = 4$.

Recall that higher dimensions may correspond to a variation of property like, temperature, pressure, color, concentration and so on.

The 5D case, will add a supplementary stage for time. By doubling, the linear time became a planar time. Figure 1.8b outlines a cycle of hyper-lines in plane that is a square of cubes. The dimension for Fig. 1.8b is $2 + 3 = 5$. The 5D object refers to a 3D cubic object that evolves in planar time. Cyclical time is an example of planar time.

The 6D situations will add a supplementary stage for time. The spatial time may be a possible physical interpretation in this case. Figure 1.8c outlines a new cycle, and a cube of cubes resulting as a kind of parallel universe model. The dimension for Fig. 1.8c is $3 + 3 = 6$.

The front and the back faces of the 6D cube correspond to the two parallel universes. The 6D object refers to a 3D object that varies in spatial time. The couple of two-cyclical time offers an example of spatial time. Multi-cyclical time allows modeling the transitions between several parallel universes. It is an example of polytopic time.

Figure 1.8d shows the 7D case and is obtained by doubling of (c) frame from Fig. 1.8. That (c) is a 6D. The dimension for Fig. 1.8d is $1 + 6 = 7$.

Figure 1.8e shows the 8D case obtained by doubling of Fig. 1.8d, a 7D frame. The dimension for Fig. 1.8e is $2 + 6 = 8$.

Figure 1.8f shows the 9D case obtained by doubling of Fig. 1.8e an 8D frame. The dimension for Fig. 1.8f is $3 + 6 = 9$.

We observe that subsequent the Polytopic Roadmaps development stages, the concepts of time advances by doubling and relating.

1.2.4 Periodic Tables

Chemists studying the material level of reality make use of theoretical tools as the periodic table for classification and quantum chemistry. Additionally the chemists have invented many devices and technologies to combine substances and afterwards to decompose and separate them.

The challenge for researchers and entrepreneurs focusing on other levels of reality, as industrial, socio-economical, logical systems is to develop analogous methods, techniques and models for coding and classifying, for composing and decomposing specific components and compounds.

Components of Polytopic Roadmaps may be organized and displayed as Periodic Tables.

Figure 1.9 shows the process of doubling and relating as a Periodic Table.

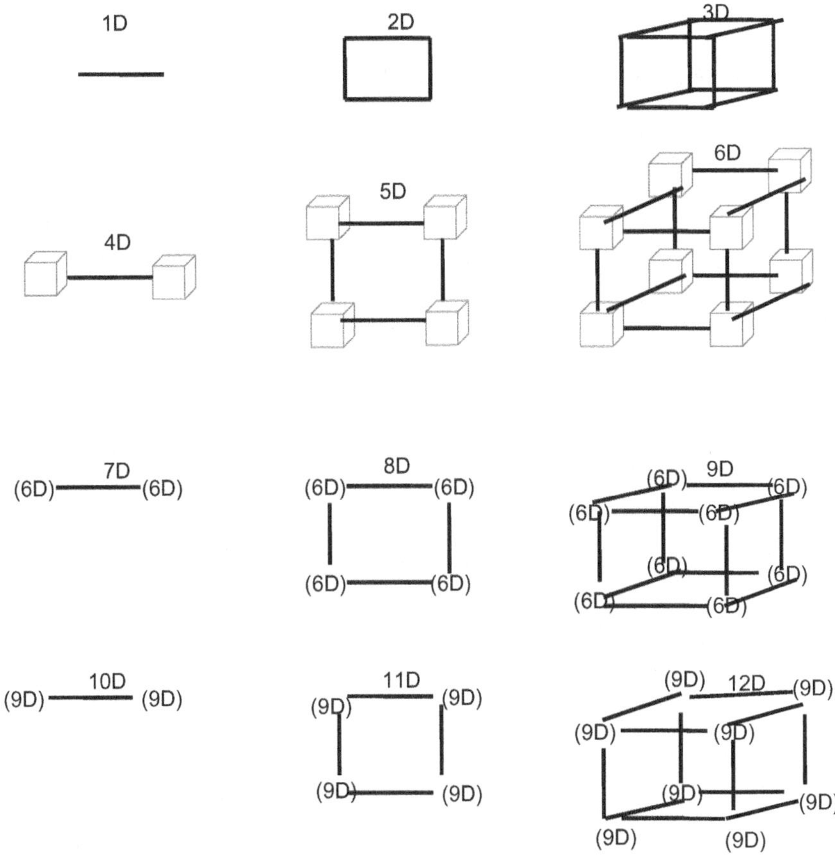

Fig. 1.9 Periodic table: doubling and relating

This table may help to illustrate the geometry of the frameworks and the concepts of time. It is also useful as a guideline towards higher dimensionality frames [2] for classification purposes, for analysis and synthesis.

The Periodic Table contains columns called groups and rows called periods.

We have three columns in the table shown in Fig. 1.9.

The columns correspond in succession to lines, squares and cubes that is to 1D, 2D and 3D. These columns highlight the line, the square and the cube as patterns or skeletons. These are the basic patterns of the Periodic Table.

Any row shows the same number of components for the periods.

The next row, unincluded in Fig. 1.9 will show 13D, 14D and 15D frames.

Rows or coupled rows of the Periodic Table generate Polytopic Roadmaps.

It is possible to range the components of Polytopic Roadmaps in different Periodic Tables.

Figure 1.10 shows the variant-1 of presentation of the Periodic Table. This contains

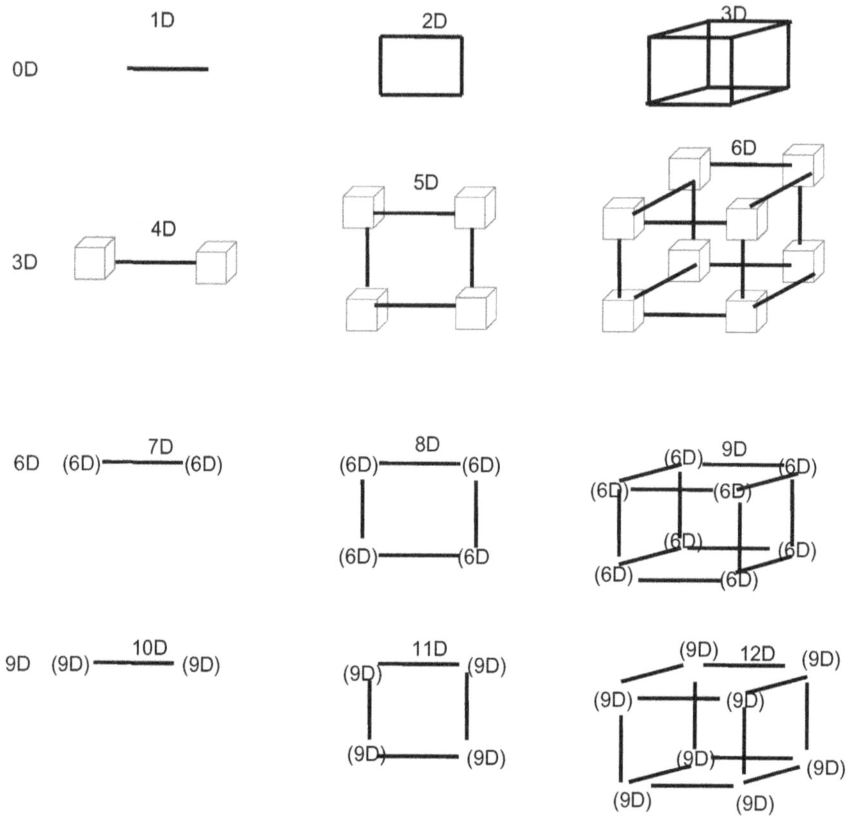

Fig. 1.10 Periodic table: doubling and relating-variant 1

a supplementary first column. We have four columns in this variant.

The columns highlight the dot, the line, the square, and the cube as pattern or skeletons.

In this case the supplementary column repeats the elements of the last column from Fig. 1.9. It shows dots 0D, cubes 3D, 6D-cubes and 9D-cubes. Theses are the building bricks for the Periodic Table, for the corresponding rows.

The next row, unincluded in this figure will show 12D, 13D, 14D and 15D frames.

Figure 1.11 shows the variant-2 of presentation of the Periodic Table.

The Periodic Table shown in Fig. 1.11 introduces a supplementary column as the last column of the table. Instead of four columns we have five columns in this variant.

The columns correspond in succession to dots, lines, squares, cubes and 4D-hypercubes that is to 0D, 1D, 2D, 3D and 4D frames. The columns highlight the dot, the line, the square, the cube and the 4D-hypercube as patterns or skeletons.

The rows use as basic bricks the 0D (dots) in the first row, 4D-cubes in the second row and 8D (hypercube of hypercubes) in the third row. The selected bricks are the final stage in the previous rows.

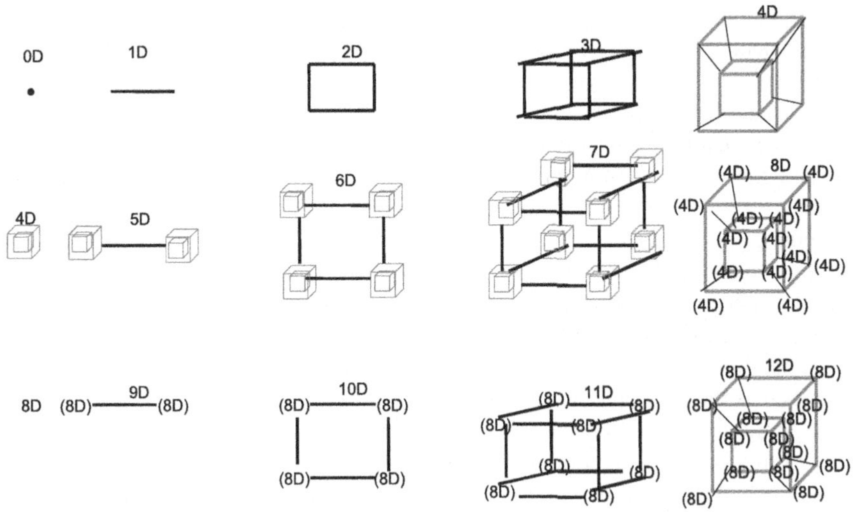

Fig. 1.11 Periodic table: doubling and relating-variant 2

Any row show the same sequence as period: dot, line, square, cube, hypercube.

The next row, unincluded in this figure will show 12D, 13D, 14D, 15D and 16D frames.

1.3 Everything Polytopic

The manufacturing sectors, are attempting to secure their market share through innovations toward the fourth industrial revolution and beyond. Industry 4.0 exploits existing advances in information technology, communications, and automation to shape a new industrial era. The goal is to create industrial sectors with the ability to compete in the global market by creating high value-added products through the innovation of products and services. Innovations will empower companies by giving them a uniquely competitive advantage of increased efficiency, resource utilization and responsiveness to the needs of both customers and society.

In a world where climate change, decreasing resources and irregular demographics are all major concerns, it is evident that the industry only will not solve all the critical problems. When it comes to industrial production, now is the time for structuring and destructuring and industry 4.0 implementations will not be enough. It is a need to advance beyond the industry 4.0 towards industry 5.0 and polytopic industry.

Personalization and mass-customization are just two examples of how customer demands are fundamentally changing. New industrial paradigms need to be created where not only environmental sustainability is protected, but also new business models that secure long-term competitiveness.

Even though industry 4.0 and society 4.0 are in their infancy, engineers and entrepreneurs are already looking beyond it, to industry 5.0 to society 5.0 and so on to the forthcoming polytopic industry and polytopic society.

A decade has passed since talk of industry 4.0 first surfaced in manufacturing circles, yet visionaries are already forecasting the next stages of revolution- industry 5.0 and more general polytopic industry. If the current industry 4.0 revolution emphasizes the transformation of factories into smart facilities that utilize cognitive computing and interconnect via cloud servers, industry 5.0 is set to focus on the return of human hands and minds into the industrial framework. Industry 5.0 is the revolution in which man and machine reconcile and find ways to work and develop together to ensure personalization and individualization to improve the means and efficiency of production. The correlation of human and machines endorses new dimensions described as the polytopic project of industry 5.0.

Everything will be connected: every asset, material, supplier, worker, warehouse, delivery truck, stakeholder, and customer. In the same time, products are rapidly becoming more intelligent and interactive, which creates new opportunities to apply emerging technologies in ways only imaginable before.

Three developments are pushing industry 4.0 to the next dimensions of polytopic industry as in the following examples:

- Intelligent digital products: Companies can now develop intelligent, connected, and self-aware products capable of sharing information about wear and tear, usage level, and storage conditions-as well as data on the health, location, or ambiance of the customer.
- Emerging technologies: New technologies are reaching a tipping point in terms of readiness and availability. Some of these include cognitive computing, 3D and 4D printing, augmented reality, blockchain, machine learning, artificial intelligence, voice-controlled user interfaces, robots and cobots.
- Shifts in customer attitudes, behaviors, and expectations: From wearable such as networked running shoes to connected cars and industrial products.

The Internet of Things are positioned to give companies and organizations even more opportunities to collect data and use it to improve customer outcomes.

These developments are making it possible to bring the consumer into the product value chain as an indispensable user contributor.

Looking ahead to polytopic industry, we can expect the further restructuring of product development and production in ways that will redefine not only manufacturing processes but also what a product is and the value it provides to customers.

As more products incorporate digital technologies, they will move from being a collection of functionalities and become digital platforms for value creation-both for the user and the creator of products.

We will begin to see tightly integrated ecosystems of companies in different industries working together to provide distinctive value to customers. This will be achieved through the creation and delivery of more complete offerings, thanks to pervasiveness and exponential growth of data and the software that uses it and to increased standardization of application interfaces, integration points, and automation technologies, which reduce conflicts and simplify the creation.

We expect polytopic industry will create new and more agile offerings centered on customers' demands, likes, needs, and preferences.

As more products become platforms for innovation and value creation, opportunities will open for companies to become digital platform adopters that drive innovation in their own right because no single company can own innovation on digital platforms.

Besides manufacturing processes or factory automation, industrial revolutions have an impact in domains such as education, healthcare, agriculture, economy and so on. So far, polytopic society will uses the industry as basis towards a human machine symbiosis throughout the society. It targets to improve the life of every person.

Various twenty-first century challenges should be addressed: climate change, food security, the limited availability of natural resources and sustainable development, the ageing population. The goal is an inclusive economy and society in which everyone shares the same opportunities and that guarantees work, economic growth and well-being for all.

References

1. Ammon, J.: Hypercube connectivity within ccNUMA architectures. SGI Origin Team, Mountain View, CA (1998)
2. Anderl, R, Picard, A., Wang, Y., Fleischer, J., Dosch, S., Klee, B., Bauer, J.: "Guideline industry 4.0—guiding principles for the implementation of Industry 4.0 in small and medium sized businesses." In: VDMA Forum Industry 4.0, Frankfurt (2015)
3. Choudhury, N.: World wide web and its journey from web 1.0 to web 4.0. Int. J. Comput. Sc. Inf. Tech. **5**(6), 8096–8100 (2014)
4. Coxeter, H.S.M.: Regular Polytopes, 3rd edn. Dover, New York (1973)
5. Helbing, D.: What the digital revolution means to us. Science Business, June 12 (2014)
6. Helbing, D.: Societal, economic, ethical and legal challenges of the digital revolution: From big data to deep learning, artificial intelligence, and manipulative technologies. Artific. Intell. Manipulative Technol. April 14 (2015)
7. Iordache, O.: Self-evolvable Systems. Machine Learning in Social Media. Springer, Berlin, Heidelberg (2012)
8. Iordache, O.: Polytope Projects. Taylor & Francis, CRC Press, Boca Raton, USA (2013)
9. Iordache, O.: Implementing Polytope Projects for Smart Systems. Springer, Cham, Switzerland (2017)
10. Iordache, O.: Advanced Polytopic Projects. Springer, Cham, Switzerland (2018)
11. Kagermann, H., Wahlster, W., Helbig, J. :(eds.) Recommendations for implementing the strategic initiative Industrie 4.0: final report of the Industrie 4.0 Working Group (2013)
12. Scharmer, O.: Theory U: Leading from the Future as It Emerges. Berrett-Koehler, San Francisco (2009)
13. Scharmer, O., Kaufer, K.: Leading from the Emerging Future: From Ego-System to Eco-System Economies. Berrett-Koehler, San Francisco (2013)
14. Schwab, K.: The Fourth Industrial Revolution. World Economic Forum, Geneva (2016)
15. Sparling, G.A.J: Germ of a synthesis: space-time is spinorial, extra dimensions are time-like. Proc. R. Soc. Lond. Ser. A Math. Phys. Eng. Sci. **463**, 1665–1679 (2007)

16. Thiel, T.: The Design of the connection machine. Des. Issues, **10**(1), 5–18, Spring (1994)
17. Tibbits, S.: 4D Printing: multi-material shape change. Architec. Des. **84**(1), 116–121 (2014)

Chapter 2
Stages

2.1 Cognitive Intelligent Roadmaps

2.1.1 Higher Dimensionality by Iteration

Higher dimensionality and cyclicality characterizes the Polytopic Roadmaps, (PTR), proposed as guideline for projects implementation in the domain of high complexity industrial systems.

It was observed that complex systems exhibit hierarchical self-organization in multiple conditioning levels and scales under constraints. Self-organization will occur when individual independent parts in a complex system interact and cooperate, in order to generate higher dimensional organizations.

The multiplicity of conditioning levels and scales explains why we make use of polystochastic models, (PSM), and Polytopic Roadmaps, PTR, for complex engineering projects implementation.

PSM describe processes of processes. It is an iteration that accompanies transition towards higher dimensionality. PSM elaborate on the iterative construction of processes of processes over processes and so on. The resulting compound processes, at different levels of conditioning and iteration show specific properties [21].

PTR, describe the Polytopic Frameworks allowing facing complexity as encountered in PSM advances towards higher dimensionality.

The starting point for PSM has been the learning models [24]. The learning is the early capability for equipments, technologies and industries committed to challenge the human behavior.

In a simplified form, the models for learning and adaptability considered only two levels or two spaces, the space of states, S and the space of conditions, K and their interaction operator, u. The learning system was presented with a series of conditions k in K, on each of which it changes states s in S, allowing sequential adaptability.

Let (s_n) in S define the states of the moment n, and (k_n) in K the conditions at the moment n. Equation (2.1), with u an operator, describes the iteration of states.

O. Iordache, *Polytopic Roadmaps*, Studies in Systems, Decision and Control 368 ,
https://doi.org/10.1007/978-3-030-75630-7_2

$$s_{n+1} = u\,(s_n,\ k_n) \tag{2.1}$$

In PSM, the space of conditions K was enhanced to multiple levels of conditioning, involving conditions K1, meta-conditions K2, meta-meta conditions K3 and so on.

The meta-condition is a condition of conditions and so on.

The usual notations are: S-states (denoted also K0), K1-processes, K2-process of processes (meta-processes), and K3-process of processes of processes (meta-meta-processes).

The iteration scenario do not limits to processes. Table 2.1 shows more examples of iterations.

Abbreviations are: EDoE-evolvable designs of experiments [20], GST-general systems theory [43], HC-hypercycles [13], HS-hyperstructures [1], MES-memory evolutive systems [12], MOF-metal organic frameworks [46], PSM—polystochastic models [21], RIFS-random iterated function systems [6], RSCC-random systems

Table 2.1 Iterations

Iteration definition	Perspectives
Categories of categories	MES
Conditions of conditions	Meta-conditions
Cycles of cycles	Hypercycles, HC
Designs of designs	EDoE
Economies of economies	Polytopic economy
Equipments of equipments	Polytopic equipments
Flowsheets of flowsheets	Polytopic flowsheets
Frameworks of frameworks	Polytopic frameworks
Functions of functions	Fractals, RIFS
Graphs of graphs	Hypergraphs
Industries of industries	Polytopic industry
Materials of materials	Polytopic materials
Models of models	Polytopic models
Molecules of molecules	Supramolecules, SM
Networks of networks	Hypernetworks
Operations of operations	Polytopic operations
Processes of processes	PSM
Societies of societies	Polytopic society
States of states	RSCC, RIFS
Structures of structures	Hyperstructures, HS
SM of SMs, MOF of MOFs	Polytopic chemistry
Systems of systems	SoS, GST
Technologies of technologies	Polytopic technologies

with complete connections [24], SM-supramolecules [29], SoS-systems of systems [32].

Figures 2.1 and 2.2 illustrate the geometrical iteration.

Figure 2.1 is presented as a Periodic Table.

The successive columns correspond to dots, 2D and 4D frames. The dimension increases from nD to (n + 2) D.

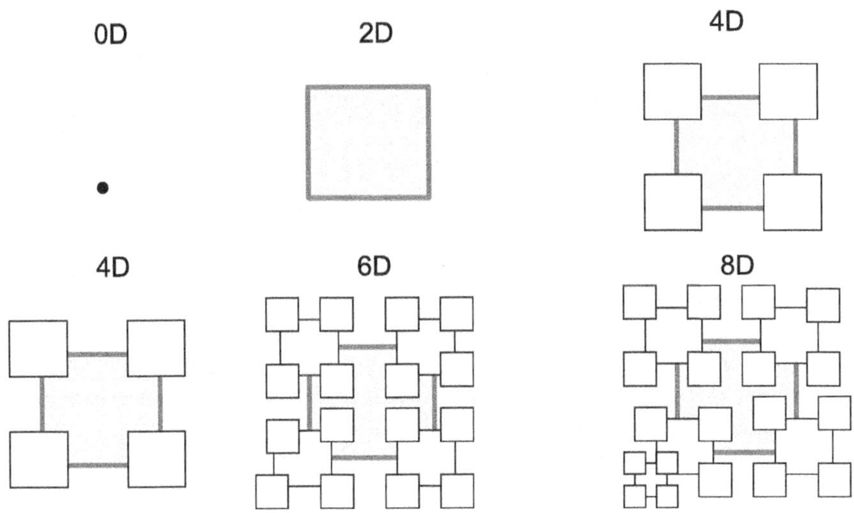

Fig. 2.1 Periodic Table: 2D iteration

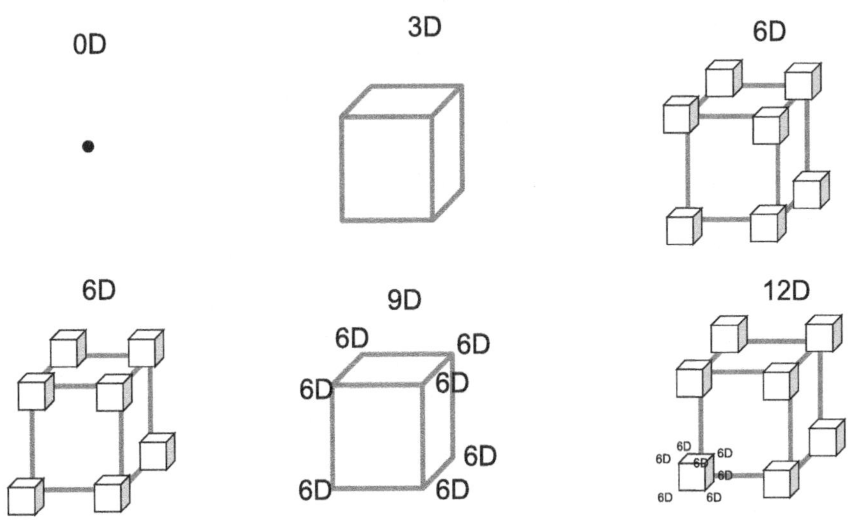

Fig. 2.2 PeriodicTable for 3D iteration

In the first row, by iterating the 0D dot, became the 2D square and this became the 4D square that is a square of squares. The dimension is $2 + 2 = 4$.

In the next row the 4D square, became the 6D square that is a square of squares of squares. The new dimension is $2 + 2 + 2 = 6$. The process continues to 8D partially represented to avoid complicated drawings.

The next row unincluded in this figure will show 8D, 10D and 12D frames.

Figure 2.2 shows the Periodic Table obtained by 3D iteration.

The first row of Fig. 2.2 shows that by iterating the 0D dot became a 3D cube and this became the 6D cube that is a cube of cubes. The dimension is $3 + 3 = 6$.

The second row of Fig. 2.2 shows that by iterating the 6D dot became a 9D cube and this became the 12D cube that is a cube of cubes of 6D frames. The dimension is $3 + 3 + 6 = 12$.

The dimension increases from nD to (n + 3) D.

Notable theoretical perspectives more or less close to polystochastic models and Polytopic Roadmaps approaches are the cybernetics [45], GST-General Systems Theory [43], HC-hypercycles [13], HS-hyperstructures [1–3] the MES—memory evolutive systems [12] and so on.

These research directions develop, in different ways, the idea of iterative constructions of systems of systems over systems and so on. The systems of different levels of reality will show specific properties. This means that we have to consider iterations in new contexts and not just a recursive procedure.

The concept of Systems-of-Systems, SoS, has been developed to characterize large, distributed, systems that consists of interacting and networked, but partially autonomous, components and can show emergent behavior. System of systems, SoS is a collection of task-oriented or dedicated systems that pool their resources and capabilities together to create a new, more complex system which offers more functionality and performance than simply the sum of the constituent systems. The methodology for defining, abstracting, modeling, and analyzing system of systems problems is typically referred to as system of systems engineering.

Baas et al. [4] compared two different approaches to the modeling of complex natural systems, in particular of their hierarchical organization with higher-order structures and their emergence processes. These approaches are the hyperstructures, HS and the memory evolutive systems, MES. The hyperstructures are structural while memory evolutive systems based on category theory, take more into account the dynamics. The dynamical organization and mechanisms developed for memory evolutive systems rely on general ideas that might be disengaged from the categorical setting and extended to the general frame of hyperstructures.

2.1.2 Closure Around the Center

As well as the higher dimensionality, the cyclical character is mandatory for the complex engineering systems. The control of such systems is accomplished by closing the dual forward and backward procedures for the control loop. Request for cyclical control explains why we need to use concepts as duality, closure, cyclical time, center, rhythms, and wave equation [22, 23]. For cyber-physical systems, CPS, the physical side endorses higher dimensionality while cyber side induces cyclicality. Iterating focuses on dimensionality while relating of the iterated parts focuses on cyclicality. The digital twin, DT is another concept associated with cyber–physical integration. The DT creates high-fidelity virtual models of physical objects in virtual space in order to simulate their behaviors in the real world and provide feedback [42].

Dual or complementary pairs are those things, events and processes in nature that may appear to be contraries but are mutually related and inextricably connected [26]. Both aspects of a complementary pair are required for an exhaustive account of phenomena. Dual aspects are dynamic, contextual, dialogical and relational.

Complementarities manifest in the whole-part behavior of reality, in the energy-informational properties of reality.

Duality studies reveal upward and downward, forward and backward, divergence and convergence, composing and decomposing, construction and deconstruction, conditioning and deconditioning and other pairs of dual processes.

Closure concepts play a major role in systems theory where may be used to identify the whole system in correlation with its environment and to define the autonomy of the systems. Different closure concepts are linked to different facets of complexity.

Centers endorse the interchange of dual parts, to ensure coordination and coherence of the evolution on different levels.

The centers concept is correlated with other concepts as "included middle", "hidden third" [30, 33, 34], "fold" [11], "middle kingdom" [27], "tilde" [26] and "iterated loop space" [31].

Following Lupasco, Nicolescu takes a transdisciplinary approach to nature and knowledge. Transdisciplinary knowledge, as based on the logic of the "included middle", necessarily leads to the inclusion of values and systemic understanding versus simple analytic reasoning. According to this logic, in every relation involving two separate levels of experience, there is a third level that belongs simultaneously to both. It is the support of relation.

If meta-theoretical background assumptions establish the framework for the disembodied mind, then other meta-theoretical assumptions must be found as a ground for framing an understanding of embodiment in which the lived body and the physical, the subjective and the objective, are complements, and not competing alternatives forever segregated into pure forms of reality. Such a set of assumptions is described from the post-Cartesian perspective termed relationism [27, 35, 36].

Latour described this perspective as a move, a way from the extremes of Cartesian split towards a center or "middle kingdom" position where all objects of knowledge are represented not as pure forms, but as forms that flow into each other across fluid boundaries. From this perspective, mind does not cause body, nor does body cause mind, nor do two entities interact but rather, mind and body are co-constituted, and as such, form non-dissociable complements [37]. This perspective is significant for cyber-physical systems, digital twins and implicitly for industry 4.0.

For Polytopic Roadmaps the "Self" plays the role of the center able to focus and coordinate the multi-level frames and initiate dimensionality increasing [22].

Modulated by "Self" , the swinging rhythms "up" and "down", forward and backward, separating and integrating, composing and decomposing, constructing and deconstructing, dialoging between the two complementary or dual ways are essential for complex projects implementation.

2.1.3 Functional Circles

The cyclical behavior and closure around the center is illustrated by the *Umwelt* concept introduced by von Uexküll in theoretical biology to describe how cognitive organisms perceive and interpret their environments. The *Umwelt* was defined as the part of the environment that an organism selects with its specific sense organs according to its needs [44]. Pursuing Kantian viewpoint, *Umwelt* theory asserts that a complex system doesn't responds to its environment but rather to its perception of the environment. A complex system actively creates its *Umwelt*, through repeated interactions with the environment. It simultaneously observes the world and changes it, the phenomenon which von Uexküll called a functional circle.

Figure 2.3 shows the basic elements of the functional circle.

Fig. 2.3 Functional circle

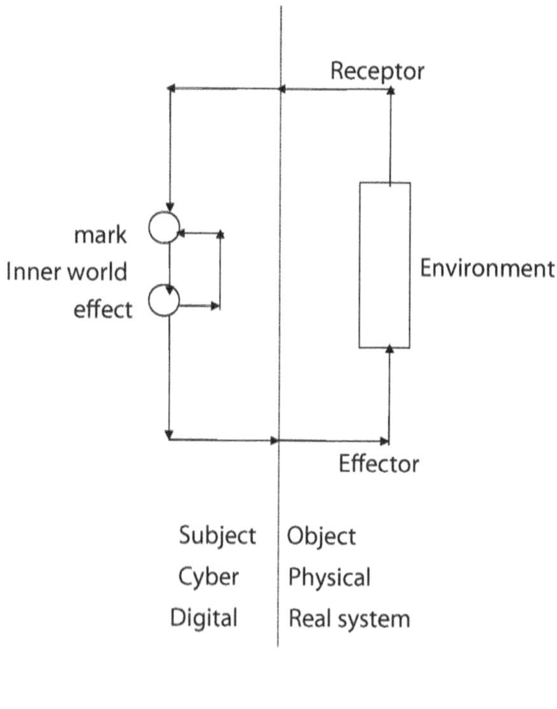

The functional circle is a simple cognitive framework that self-organizes in order to accommodate the environment. The functional circle includes receptors and effectors. The sensory experience is based on interactions and these have specific purposes. The elementary unit of self-evolvable systems includes the functional circle of the four parts: environmental objects, receptors, inner world and effectors.

Figure 2.3 outlines the two-world perspective, the outer and the inner world. Their specific interaction endorses the higher dimensionality.

The inner world includes a "mark" organ and an "effect" organ.

Figure 2.3 illustrate the relation of the functional cycle with modern concepts as cyber-physical systems [5, 28] and digital twins [25, 42]. Typically described as consisting of a physical entity, a virtual counterpart, and the data connections in between, the digital twin is increasingly being explored as a mean of improving the performance of physical entities through leveraging computational techniques, themselves enabled through the virtual counterpart.

In *Umwelt* studies we are confronted with a digital twin biosystem.

The external world corresponds to the physical part and to real system.

The inner world corresponds to the cyber part and to the virtual part of digital twin.

2.1.4 Action Schemes

Circular reactions have been outlined by the study of action schemes done by Piaget [38, 39]. Piaget called his general theoretical framework genetic epistemology since he was primarily interested to study how knowledge develops in living organisms. Cognitive structures are patterns of physical or mental actions that underlie specific acts of intelligence and correspond to the stages of development.

The Piaget's action scheme, which constitutes the foundation of his learning theory, is a cycle including three elements: a recognized situation, an activity that has been associated with this situation, and an expected result. The recognition of a situation involves assimilation, that is to say, the situation must manifest certain characteristics which the organism has abstracted in the course of prior experience. The recognition then triggers the associated activity. If the expected result does not occur, the organism's equilibrium is disturbed and an accommodation may occur, which may eventually lead to the formation of a new action scheme. Accommodation does not take place unless something unexpected happens. Assimilation integrates new information in pre-existing structures while accommodation change and build new structure to understand new information. Piaget general equilibration theory supports the cognitive-intelligent inspiration for engineering.

Figure 2.4 shows a polytope associated to the action scheme.

The direct sequence S → K1 → K2 → K3 corresponds to assimilation stages. The reverse sequence K3' → K2' → K1' → S' corresponds to accommodations stages.

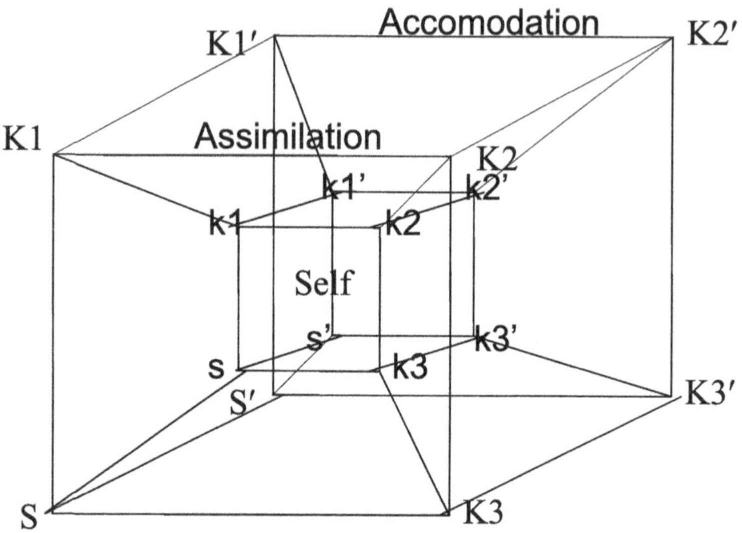

Fig. 2.4 Polytope 4D for action scheme

After a step in the direct sequence, the systems contact the "Self" represented by the inner cube. Signals as s, k1, k2 and k3 reinforce the evolution according the direct sequence S → K1 → K2 → K3. The set (s, k1, k2, k3) corresponds to assimilation.

If the "Self" looks to change the direction shifting to k3′, k2′.k1′, s′ the next step will be along the reverse sequence: K3' → K2' → K1' → S'. The set (s′, k1′, k2′, k3′) corresponds to the accommodation.

2.1.5 Piagetian Developmental Stages

Information-processing theories have yielded measures of cognitive complexity, which can be considered a major organizing theme in cognitive-intelligent development [18].

An early approach is represented by the Piagetian models, which sought to explain the course of cognitive development, in terms of the growth of information processing capability.

Cognitive and intelligent structures are patterns of physical or mental actions that underlie specific acts of intelligence and correspond to the stages of development [38, 39].

According to Piaget, there are four cognitive-intelligent development stages:

- Sensory-Motor
- Preoperational
- Concrete Operational

- Formal.

It was observed that restriction of cognitive-intelligent capability to the formal stage may correspond to systems stagnation and possible failure for evolving automata [7]. This refers to automata that have a code or protocol that recommend some actions for situations requiring a different code.

Growing complexity impose to transcend learning and to look forward for creativity, self-evolvability and smartness or intelligence for operations and automata (Iordache 2012).

Piaget's epistemology opens the road for cognition beyond the fourth stage. Piaget initiated the study of post formal stages, in which agents are able to operate over logical systems. This refers to post-processing of logic and formal theories expressed in that logic. Elaboration and modification of axiomatic schemes may be considered as surpassing the formal stage and are to formal schemes what the latter are to concrete operations [40, 41].

The post formal stages appeared as possible candidates for the so-called 5th cognitive development stage [7]. They are comparable to the framework in which post formal reasoning involves the "Self".

Figure 2.5 shows a 4D polytopic presentation of the Piagetian developmental stages.

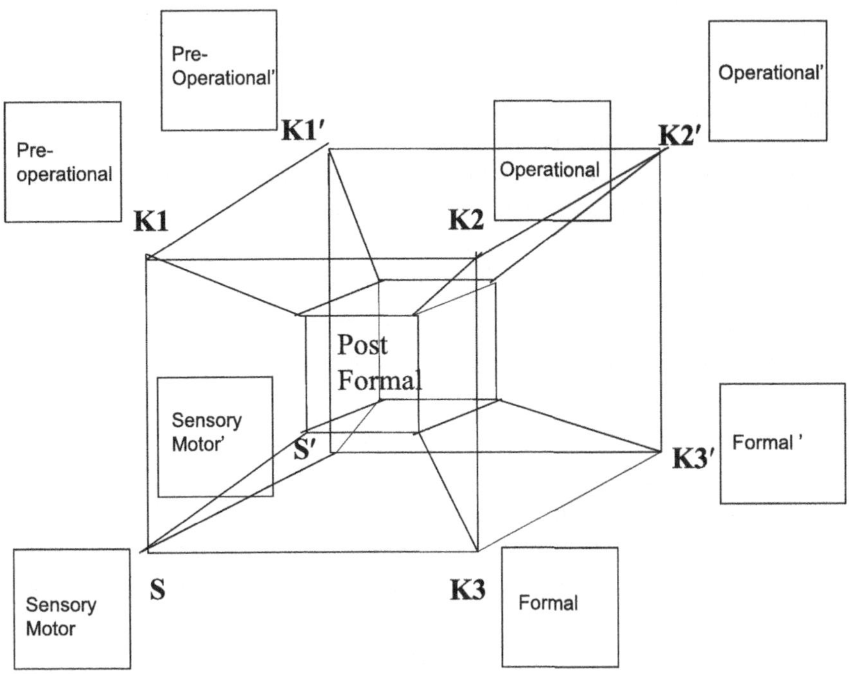

Fig. 2.5 Polytope 4D for Piagetian development stages

The four stages of Piaget, associated to S, K1, K2 and K3, have been completed in Fig. 2.5 by the self-evolvability stages S′, K1′, K2′ and K3′. This allows describing systems able to self-evolve by internal structures modification.

There are four stages on the front face of the polytope. The notations are:

- S-Sensory Motor
- K1-Preoperational
- K2-Operational
- K3-Formal.

The development is represented clockwise.

The central stage the "Self", may ensure the cooperation and redistribution of the four stages to another face of the polytope, with another starting stage.

Any stage embeds the previous ones. After one cycle an augmented reality may support a new cycle of development. The post formal stages appear as a working example of cognitive "Self".

As shown in Fig. 2.5 two ways should be considered for development.

This means that complementing the direct, construction way S → K1 → K2 → K3 we need to look at the reverse, deconstruction way K3' → K2' → K1' → S'.

We can identify:

- S′-Sensory Motor′
- K1′-Preoperational′
- K2′-Operational′
- K3′-Formal′.

The new stages are modified, possibly improved.

Through the developments of the direct way, it is possible to modify results, improve cognitive capabilities for the reverse way. The swinging from direct to modified developmental stages mediated by the "Self" is the source of creativity in complex problem solving or technology development.

That is because the boundaries where creative research stand out, the old information is evaluated and new information is created, consist of coexisting tendencies. Integration and differentiation, construction and deconstruction coexist and the meta-stable coordination dynamics emerges as a specific blend of integration and differentiation tendencies.

2.2 Dynamic Skill Theory

The development stages theory, of Fischer and coworkers is based on cognitive skills, where skills refer to control over sources of variation in a person's own behavior [14, 15, 17]

It is considered that there are several major developmental stages or tiers:

- Sensory motor actions

- Representations
- Abstract
- Principles.

These can be compared to Piagetian four stages. The stages have similarities but do not coincide. Notice that a supplementary major stage of reflexes, before sensory motor actions stage has been also considered in some preliminary studies.

The major stages of DST are structured in sub-stages.

This corresponds to a significant insight of neo-Piagetian theory if compared to Piaget theory. It draws attention to a fine structure for development.

Within each major stage or tier there is a recurring cycle of four sub-stages:

- single set
- mapping (of sets)
- system (composition of mappings)
- system of systems.

To these sub-stages we associate the labels, 0, 1, 2 and 3. They may be considered as low dimensional part of Polytopic Roadmaps.

Figure 2.6 illustrates the Polytopic Roadmaps of sub-stages.

These Polytopic Roadmaps with single set, mapping, system, and system of systems, (SoS), have a general validity, for different levels in complex systems studies [14, 16].

The possibility of a fifth sub-stage starts to be discussed in industry 4.0 and society 5.0 studies. This fifth sub-stage could correspond to systems of systems of systems, SoSoS and may be illustrated by a 4D hypercube completing Fig. 2.6 sequence.

In some cases the highest sub-stage of one major stage is shared with the lowest sub-stage of the next, and represents a transition between major stages.

It was supposed that each new major stage represents a new dimension and this is marked by a new behavioral control system, which is supported by a new kind of neural network [17]. Knowledge is built by repeated reconstructions that move

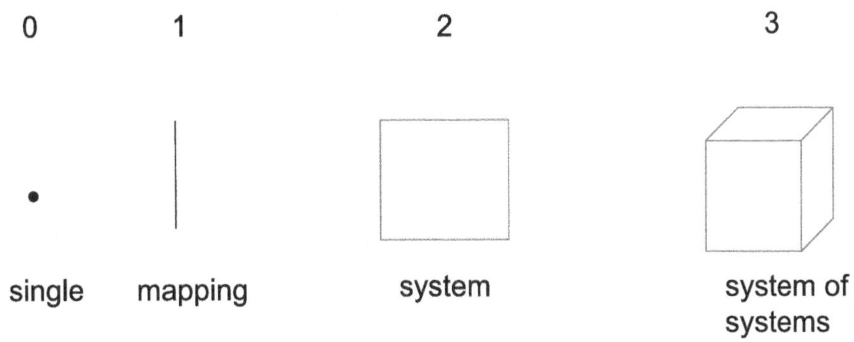

Fig. 2.6 Polytopic roadmaps of sub-stages

towards higher complexity and abstraction, though with many reversals and recoveries. The recurring cycles of development that occur in each stage are supported by empirical observations of brain growth.

The implications of the developments of dynamic skill theory are significant [18]. They offer an explanation of the anomaly, crucial to understanding development, that there is both variability and consistency in cognitive-intelligent development. Dynamic systems justify variability from a relatively small set of common processes.

The four major stages in DST considered for Polytopic Roadmaps presentation are denoted as follows:

- Sensory Motor Actions-SM
- Representations-Rep
- Abstractions-Abs
- Principles-Pri.

The four sub-stages associated to any major stage, and the modified four sub-stages are denoted:

0-single
1-mapping
2-systems
3-system of systems
0′-modified single
1′-modified mapping
2′-modified systems
3′-modified system of systems.

The modified sub-stages pertain to modified Polytopic Roadmaps.

Table 2.2 shows the connections of the developmental major stage for dynamic

Table 2.2 Developmental stage for dynamic skills

			Pri-3-SoS
			Pri-2-systems
			Pri-1-mapping
		Abs-3-SoS	Pri-0-single
		Abs-2-systems	
		Abs-1-mapping	
	Rep-3-SoS	Abs-0-single	
	Rep-2-systems		
	Rep-1-mapping		
SM-3-SoS	Rep-0-single		
SM-2-systems			
SM-1-mapping			
SM-0-single			

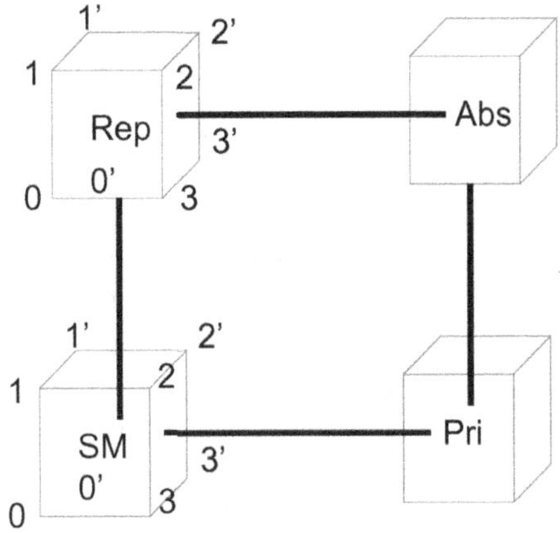

Fig. 2.7 Polytope 5D for dynamic skill development

skills.

The highest sub-stage of one stage is shared with the lowest sub-stage of the next stage.

Figure 2.7 shows the polytope 5D for dynamic skill development.

Using the thick hyper-lines, the 5D cube is shown as square of cubes.

The dimension is $2 + 3 = 5$ where 2 corresponds to the thick hyper-lines square and 3 to the cubes.

The main major stages are represented as cubes denoted: SM, Rep, Abs and Pri.

Figure 2.8 shows the polytope 6D for dynamic skill development.

It is 6D since the main stages are considered as 3D cubes. This follows the structuring of stage in sub-stages.

With the thick hyper-lines, the 6D cube is shown as a cube of cubes.

The main stages SM, Rep, Abs and Pri are modified to the main stages: SM′, Rep′, Abs′ and Pri′ to account for major stage modifications during development.

The polytopes 5D and 6D appear as superposition of Polytopic Roadmaps of sub-stages. Figure 2.8 outline two scales of periodicities, at the major level of stages and at the minor level of sub-stages.

This is an example of framework of frameworks.

A step forward for DST is proposed in what follows. It refers to internal control mechanism for each of the main four stages and for whole cognitive-intelligent framework.

Figure 2.9 shows the polytope 8D for dynamic skill development.

With the thick hyper-lines, the 8D polytope is shown as a hypercube of hypercubes.

Replacing the cubes by hypercubes corresponds to the increase of the number of sub-stages from four to five.

As the internal little cubes allow the control of the stages SM, Rep, Abs and Pri, it results that the frameworks from Fig. 2.9 will be appropriate for systems with distributed control.

Figure 2.10 shows the developmental processes described by dynamical skill

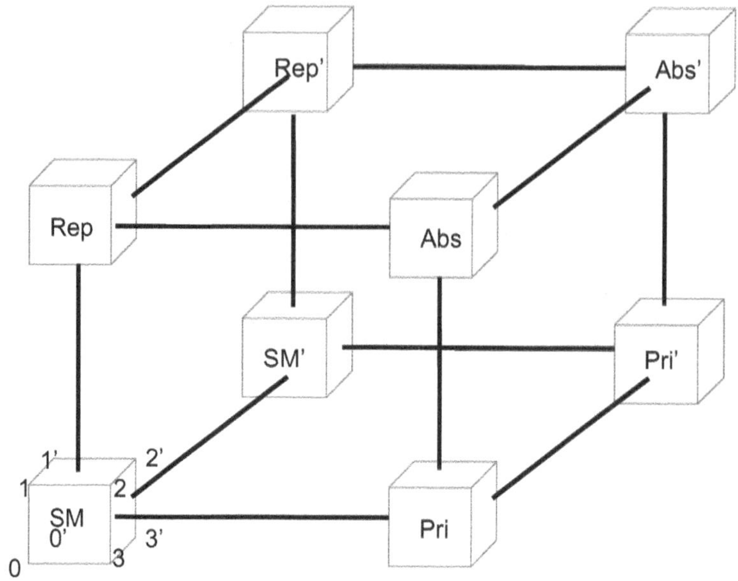

Fig. 2.8 Polytope 6D for dynamic skill development

Fig. 2.9 Polytope 8D for dynamic skill development

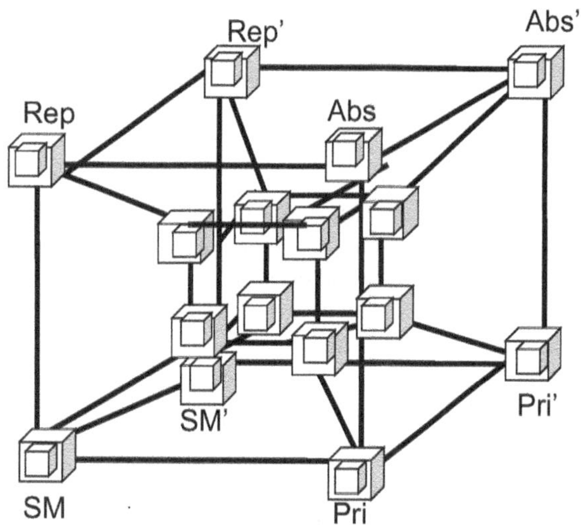

theory as a Periodic Table.

This table is useful to illustrate the developmental spiral [10]. A spiral development result by superposition of linear and cyclic processes.

The columns in the Periodic Table correspond to single set (dots), mapping (lines), systems (squares), and systems of systems (cubes) .

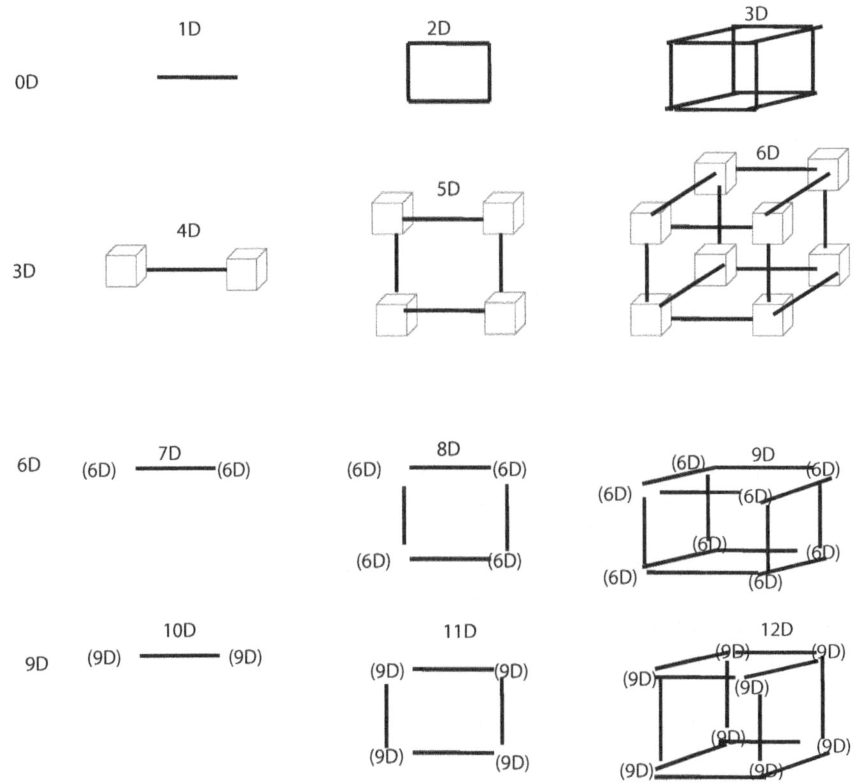

Fig. 2.10 Periodic Table for dynamic skill theory

The first row illustrates the sensory motor actions. It starts from dots and finish with 3D-cubes. The second row illustrates the representations. It starts from 3D-cubes and finish with 6D-cubes. The third row illustrates the abstractions. It starts from 6D-cubes and finish with 9D-cubes.

The fourth row illustrates the principles. It starts from 9D-cubes and finish with 12D-cubes. The fifth row, omitted in this figure, may illustrate the post formal stages. In DST they may be identified as post-principles stages. This row will start from 12D-cubes and will finish with 15D-cubes.

The rows outline the periods: dots, lines, squares, cubes. Notice that this corresponds to the identification singles, mappings, systems, system of systems in dynamic skill theory but these are just starting frames.

The first column shows the basic bricks: dots, 3D-cubes, 6D-cubes, 9D-cubes.

The second column is based on the line, the third column on the square, and the fourth column on the cube. These are the patterns or skeletons for successive columns or groups.

The next row unincluded in this figure will show 12D, 13D, 14D and 15D frames.

The Periodic Table shown in Fig. 2.10 suggests that the doubling or twining followed by relating may be the deep mechanisms for DST.

2.3 Executive Control Structures

The idea that changes in processing capacity can be described as a progression along single line of development was rejected by Case [8]. Instead, he maintained that processing capacity development recycles over a succession of four major stages and that each of them is characterized by a different kind of mental structures. These stages correspond to Piaget's main stages of sensory motor, pre-operational, concrete operational and formal operational. Each of these major four stages involves its own executive control structures that are defined by the medium of representation and the type of relations that are possible at the stage.

It should be noted that Fischer's conception of the stages of cognitive-intelligent development is very similar to that of Case. Fischer describes four major stages or tiers which have similarities with Case's major stages and Piaget's major stages.

A fine sub-structure of stages was considered by Fischer and by Case.

Executive control structures enable to represent the problem situation; to specify the objectives of problem solving and to conceive of the strategy needed to attain the objectives.

Case maintained that there are four types of executive control structures.

- Sensory motor structures from 1 to 18 months of age (perceptions and actions such as seeing)
- Inter-relational structures from 18 months to 5 years of age (mental representations that stand for actual objects in the environment, such as words or mental images)
- Dimensional structures from 5 to 11 years (mental representations that are connected together by a consistent relation such that every particular case can be related to every other case, such as the mental number line where every number can be related to every other number)
- Vectorial structures from 11 to 19 years (relations between the dimensions of the previous stage, such as ratios and proportions which connect two or more dimensions with each other).

Development within each of these four main stages evolves according to the same sequence of the four sub-stages of complexity, thus, structures of increasing complexity can be handled at each of the four sub-stages.

The four sub-stages are denoted here by 0, 1, 2 and 3 and described as follows.

- Sub-stage 0, operational consolidation (when a particular mental unit specific to each of the four main stages above can be handled, such as an action in the sensory motor stage, a word in the relational stage, a number in the dimensional stage, and so on)
- Sub-stage 1, unifocal coordination, (when two such units may be interrelated)

- Sub-stage 2, bifocal coordination, (when three such units may be interrelated)
- Sub-stage 3, elaborated coordination, (when four such units may be interrelated).

Obviously the sequence of sub-stages 0, 1, 2, 3 may be considered as elements of a Polytopic Roadmaps as that shown in Fig. 2.6 for DST.

According to Case, the expansion of the capacity of short-term storage space is caused by increasing operational efficiency. The command of the operations that define each kind of executive control structures improves, thereby freeing space for the representation of goals and objectives.

Successive stages are not unrelated. The final sub-stage of a given stage is at the same time the first sub-stage of the next stage.

Table 2.3 shows the developmental stages for staircase model.

Figure 2.11 shows the polytope 5D for executive structures development,

The four major stages of the model are denoted by SM, Int, Dim and Vec.

The four sub-stages associated to any stage, and the modified four sub-stages are denoted:

0-operational
1-unifocal
2-bifocal
3-elaborated
$0'$-modified operational
$1'$-modified unifocal
$2'$-modified bifocal
$3'$-modified elaborated.

Figure 2.12 shows the polytope 6D for executive control structure development.
The modified four major stages are denoted by SM', Int', Dim' and Vec'.

Table 2.3 Developmental stages for staircase model

			Vec-3-substage
			Vec-2-substage
			Vec-1-substage
		Dim-3-substage	Vec-0-substage
		Dim-2-substage	
		Dim-1-substage	
	Int-3-substage	Dim-0-substage	
	Int-2-substage		
	Int-1-substage		
SM-3-substage	Int-0-substage		
SM-2-substage			
SM-1-substage			
SM-0-substage			

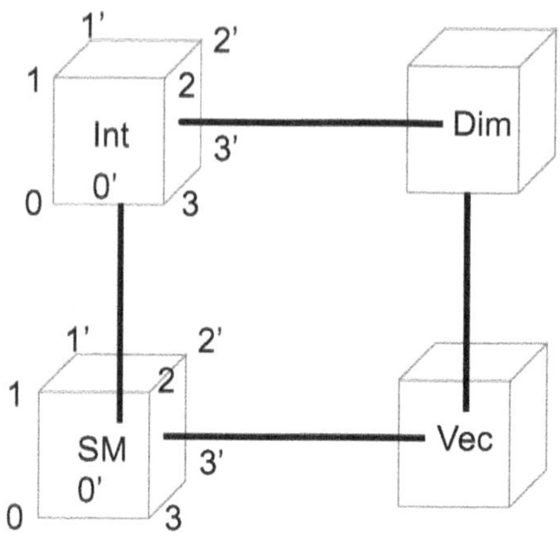

Fig. 2.11 Polytope 5D for executive control structures development

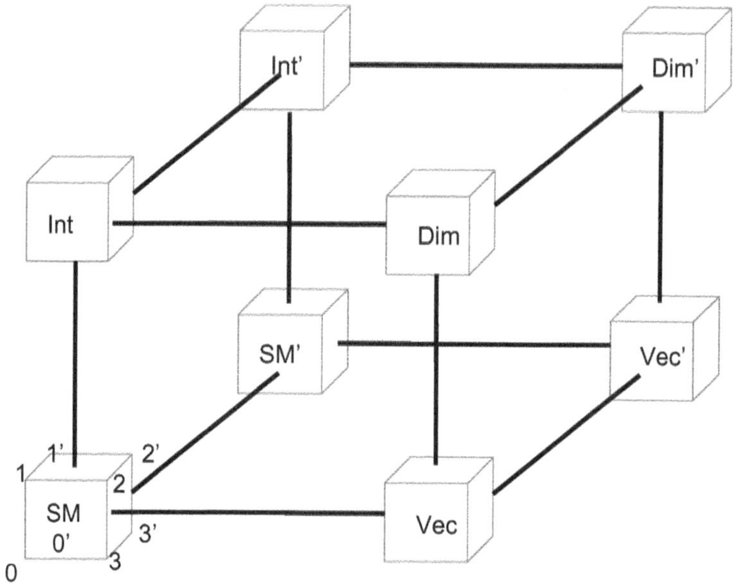

Fig. 2.12 Polytope 6D for executive control structure development

Each of the four basic stages involves its own executive control structures that are defined by the medium of representation and the type of relations that are possible at the stage.

Figure 2.13 shows the polytope 8D for executive control structure development.

As the internal little cubes allow the control of the stages SM, Int, Dim and Vec, it results that the frameworks from Fig. 2.13 will be appropriate for systems with distributed local control.

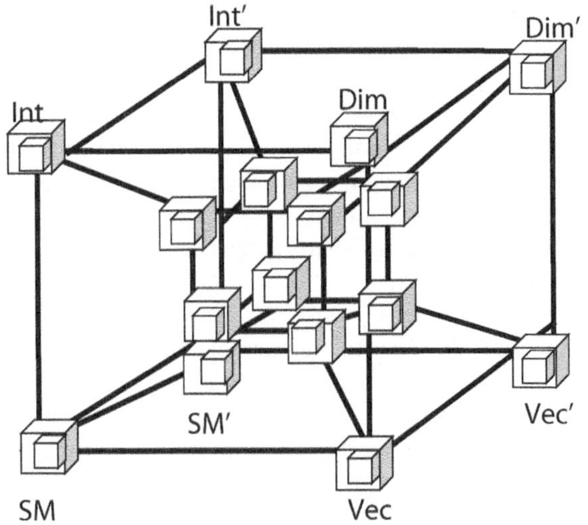

Fig. 2.13 Polytope 8D for executive control structure development

2.3.1 Hierarchical Complexity

The Model of Hierarchical Complexity, MHC, is a framework used to explore and organize the patterns of human development. It is a theory used when working with behavioral development in particular.

The MHC serves to give rank or order to the developmental complexity of a certain behavior. It is related to other hierarchical theories of development, such as that of Piaget. MHC was originally developed by Commons [9].

MHC is based on a system of task analysis in which tasks are broken down into small steps and then analyzed based on their complexity in relation to the individual's development. The task is initiated when the individual is presented with a stimulus, and their behaviors in reaction to that task are what are analyzed for their developmental complexity.

The stages of hierarchical complexity are described in the following.

- Stage 0-automatic
  Individual can respond to one environmental stimulus at a time.
- Stage 1-sensory and motor
  Individual can move their members and parts of their face and view things. They begin to react to conditioned stimuli and form responses with discrimination.
- Stage 2-circular-sensory-motor
  Individual can reach, touch, or grab things. In terms of speech, individual is beginning to grasp word parts.
- Stage 3-sensory-motor
  Individual responds to stimuli based on their established concept of it.
- Stage 4-nominal

Individual can use names for objects as well as make commands. They begin to identify relationships between the concepts they have established and can name those as well.

- Stage 5-sentential
  Individual can imitate others and acquire words or sentences through imitation, as well as following a sequence of acts presented by others.
- Stage 6-preoperational
  Individual practices early deductive reasoning engages in imitation, and can follow and tell a story.
- Stage 7-primary
  Individual begins to make simple logical reasoning with regard to rules and time, also begins to understand simple arithmetic.
- Stage 8-concrete
  Individual can engage more deeply with arithmetic problems, understand more about social and group relations, and establish relationships with others and self.
- Stage 9-abstract
  Individual begins to understand variables such as stereotypes and logic skills become more formalized.
- Stage 10-formal
  Individual can argue using linear logic can solve problems with one unknown, such as in algebra.
- Stage 11-systematic
  Individual can establish and understand systems and understands relationships and logic problems with more than one variable or unknown.
- Stage 12-metasystematic
  Individual can form relationships between systems and understand similarities and differences.
- Stage 13-paradigmatic
  Individual can fit these new meta-systems to form larger-picture paradigms, as well as point out inconsistencies between the various meta-systems.
- Stage 14-cross-paradigmatic
  Individual can see relationship between multiple paradigms.
- Stage 15-meta-cross-paradigmatic
  Individual can reflect on the similarities and differences between paradigmatic relationships.

These stages are linked to an age or developmental milestone in an individual's life in various stage models. A preliminary calculatory stage was considered before automatic stage, by some researchers. Some developmental studies suggested that the stages are less significant beyond stage 11, the systematic stage.

Neo-Piagetian perspective is opening the way to cognitive-intelligent development for post formal operations, in somewhat a stage like appearance.

Table 2.4 compares the stages in different approach for development

The neo-Piagetian theories maintain the main stages of the Piagetian theory and introduce sub-stages.

Table 2.4 Development stages theories

	Piaget	Dynamic skill theory, DST	Executive control structures, ECS	Model hierarchical complexity, MHC
K0	SM	SM Actions 0 single 1 mapping 2 systems 3 SoS	SM 0 operational 1 unifocal 2 bifocal 3 elaborated	SM 0 (0) automatic 1 (1) sensory & motor 2 (2) circular SM 3
K1	Pre-Operational	Representational 0 single 1 mapping 2 systems 3 SoS	Interelational 0 operational 1 unifocal 2 bifocal 3 elaborated	Pre-Operational 0 (3) sensory-motor 1 (4) nominal 2 (5) sentential 3
K2	Concrete-Operational	Abstract 0 single 1 mapping 2 systems 3 SoS	Dimensional 0 operational 1 unifocal 2 bifocal 3 elaborated	Concrete-Operational 0 (6) pre-operational 1 (7) primary 2 (8) concrete 3
K3	Formal	Principles 0 single 1 mapping 2 systems 3 SoS	Vectorial 0 operational 1 unifocal 2 bifocal 3 elaborated	Abstract Formal 0 (9) abstract 1 (10) formal 2 (11) systematic 3
K4	Post Formal	Post Principles 0 single 1 mapping 2 systems 3 SoS	Post Vectorial 0 operational 1 unifocal 2 bifocal 3 elaborated	Post Formal 0 (12) meta-systematic 1 (13) paradigmatic 2 (14) cross-paradigmatic 3

Notations are: SM sensory-motor, SoS systems of systems

K0–K4 are the usual notations for Piagetian stages and are associated to 4D polytope. In some studies K0 is denoted S, while K4 was assigned to "Self".

Post Principles and Post Vectorial have been identified here as Post Formal.

The MHC stages are ranked in monotone increasing order of complexity and dimensionality. Periodic properties for different stages have not been noted in MHC despite their significant role in DST or ECS theories. This corresponds to the linear classifications studies in chemistry before the arrival of Periodic Tables.

A tentative introduction of periodicity for MHC based on Table 2.4 is shown in Table 2.5. Table 2.5 refers to MHC theory only. The names of columns take into account other development theories as Piaget theory or DST. The table contains the stages as identified in MHC from 0 to 14. They have been considered as sub-stages in other development theories. As shown also in Table 2.4 the stages 0, 3, 6, 9 and 12 from MHC pertains to the column singles in DST. The stages 1, 4, 7, 10 and 13 from MHC pertain to the column mappings in DST while the stage 2, 5, 8, 11 and 14 from MHC pertain to the column systems in DST.

Table 2.5 Compact Periodic Table for model of hierarchical complexity

	Singles	Mappings	Systems
Sensory motor	0	1	2
Pre-operational	3	4	5
Concrete operational	6	7	8
Abstract formal	9	10	11
Post formal	12	13	14

It is expected that the stages in the same column outline some degree of similarity related to the associated skeletons.

For singles, this is related to dot, for mappings is related to line, while for systems is related to square. The stages from the same column have the same pattern or skeleton and this should be significant.

Compact Periodic Tables similar to Table 2.5 may be considered for other development theories.

Table 2.6 refers to DST theory only. The names of columns are based on DST.

The abbreviations are:

- Sensory Motor Actions-SM
- Representations-Rep
- Abstractions-Abs
- Principles-Pri
- Post Principles-PPri.

Post Principles have been identified here as Post Formal.

Table 2.7 refers to ECS theory only. The names of columns are based on ECS.

The abbreviations are:

- Sensory Motor-SM
- Interrelational-Int
- Dimensional-Dim
- Vectorial-Vec
- Post Vectorial-PVec.

Post Vectorial has been identified here as Post Formal.

Table 2.6 Compact Periodic Table for dynamic skill theory

	Singles	Mappings	Systems	SoS
SM	SM-singles	SM-mappings	SM-systems	SM-SoS
Rep	Rep-singles	Rep-mappings	Rep-systems	Rep-SoS
Abs	Abs-singles	Abs-mappings	Abs-systems	Abs-SoS
Pri	Pri-singles	Pri-mappings	Pri-systems	Pri-SoS
PPri	PPri-singles	PPri-mappings	PPri-systems	PPri-SoS

Table 2.7 Compact Periodic Table for executive control structures

	Operational	Unifocal	Bifocal	Elaborated
SM	SM-operational	SM-unifocal	SM-bifocal	SM-elaborated
Int	Int-operational	Int-unifocal	Int-bifocal	Int-elaborated
Dim	Dim-operational	Dim-unifocal	Dim-bifocal	Dim-elaborated
Vec	Vec-operational	Vec-unifocal	Vec-bifocal	Vec-elaborated
PVec	PVec- operational	PVec-unifocal	PVec-bifocal	PVec-elaborated

2.3.2 Strategies

The majority of the presented Polytopic Roadmaps outlined two sequences for evolution the direct sequence S → K1 → K2 → K3 that corresponds to the developmental stages and the reverse sequence K3' → K2' → K1' → S' that corresponds to the modified stages.

This is a conservative or conventional type of development. The directly acquired development is questioned and modified by the reverse sequence. The bottom-up construction from direct sequence complements the top-down deconstruction of the reverse sequence.

The conventional strategy is based on four stages K1, K2, K3 and K4. Conceivably in some cases we need to consider five stages, including the "Self" that centers the frame.

An exploratory or innovative type of development that maintains the direct and the reverse senses to ensure the global equilibrium, but radically change the stages of reverse sequence may be considered as a candidate for future investigations.

Figure 2.14 shows the exploratory polytope 4D while Fig. 2.15 shows the

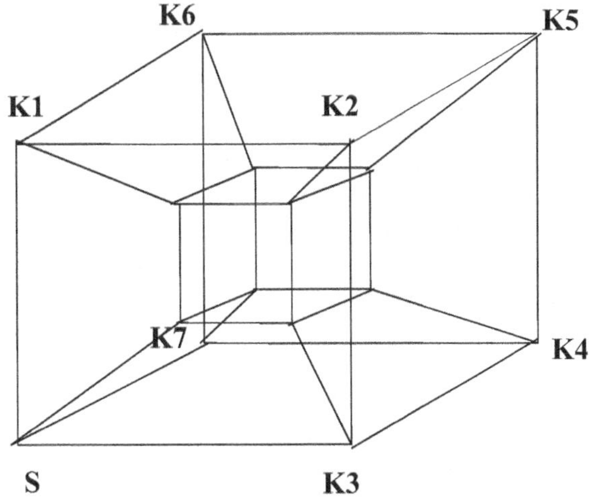

Fig. 2.14 Exploratory polytope 4D

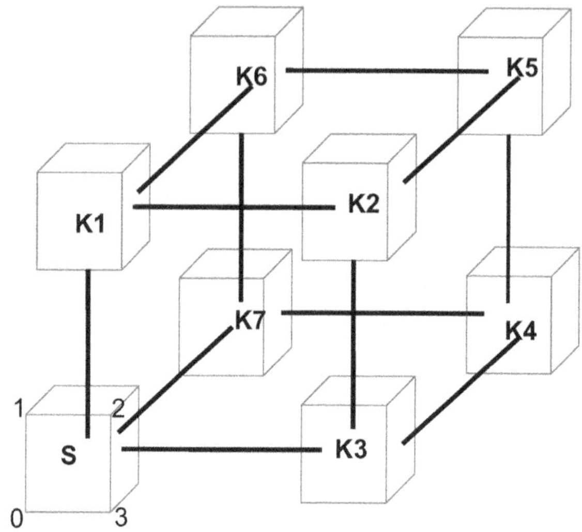

Fig. 2.15 Exploratory polytope 6D

exploratory 6D polytope.

The polytopes are 4D if they refer to stages represented as dots and 6D if they refer to major stages or tiers structured in the four sub-stages and represented as cubes.

For exploratory development we need eight different stages instead of four.

For exploratory development the direct sequence is: S → K1 → K2 → K3 while the reverse sequence will be: K4 → K5 → K6 → K7.

The reverse sequence is not just a deconstruction but a reconstruction. Reconstruction implies deconstruction but is going forward.

A continuous development involving all the stages would be: S → K1 → K2 → K3 → K4 → K5 → K6 → K7.

Finally K7 is in the same position relative to S as it was S' in conservative Roadmaps S → K1 → K2 → K3 → K3′ → K2′ → K1′ → S′ and ready for a new start.

Cycles are replaced by spirals and the center "Self" may became superfluous.

The transition from direct sequence to the reverse sequence is possible without central coordination. A procedure in which the center plays its role for conservative periods and became ineffective for exploratory periods is of interest.

Arriving in K3 the system has to choose between closing the loop establishing a center and modifying the already achieved stages according to the sequence: K3' → K2' → K1' → S' or search for an exploratory innovative sequence K4 → K5 → K6 → K7.

This mechanism is valid for major stages or tiers as shown in Fig. 2.15.

Moreover a similar strategic option conservative vs. exploratory may be considered for every sub-stage.

Table 2.8 Exploratory stages

	K4	K5	K6	K7
Post principles	PPri-singles	PPri-mappings	PPri-systems	PPri-SoS
Post vectorial	PVec-operational	PVec-unifocal	PVec-bifocal	PVec-elaborated
Post formal	Meta-systematic	Paradigmatic	Meta-Paradigmatic	Meta-cross-paradigmatic

After the bottom-up sequence of sub-stages $0 \rightarrow 1 \rightarrow 2 \rightarrow 3$ the system may search for conservative modification top-down sequence $3' \rightarrow 2' \rightarrow 1' \rightarrow 0'$ or seek an innovative exploratory sequence $4 \rightarrow 5 \rightarrow 6 \rightarrow 7$.

Certainly new mechanisms, stages and sub-stages for innovative development are waiting to be identified and explored.

A hypothesis to be considered is that the innovative stages onset would parallel the development of the post formal sub-stages in neo-Piagetian theories.

Table 2.4 identifies four sub-stages for DST, ECS and MHC.

In that table the post formal sub-stages are collected under K4 but we will suppose that the four post formal sub-stages suffer an explosion a blooming seeking for an exploratory strategy.

This explosion allows to the four post formal sub-stages to become four exploratory stages.

Table 2.8 summarizes this hypothesis.

PPri denotes Post Principles, PVec denotes Post Vectorial. Both are considered in the class of Post Formal stages.

Let us make use of the supplementary abbreviations:

- Meta-Systematic-MS
- Paradigmatic-P
- Meta-Paradigmatic–MP
- Meta-Cross-Paradigmatic-MCP.

Figure 2.16 shows the post formal exploratory 6D polytope for the MHC theory.

Obviously we can replace the sequence, MS $\rightarrow$ P $\rightarrow$ CP $\rightarrow$ MCP by the sequences:

- PPri-singles $\rightarrow$ PPri-mappings $\rightarrow$ PPri-systems $\rightarrow$ PPri-systems-of-systems.
- PVec-operational $\rightarrow$ PVec-unifocal $\rightarrow$ PVec-bifocal $\rightarrow$ PVec-elaborated.

2.4 Relational Complexity

Relational complexity model of cognitive-intelligent development is based on the idea that capacity limitations in working memory are not simply a function of the number of representational units that can be retained at any given time, but instead rejects the complexity of the relations between representational units [19].

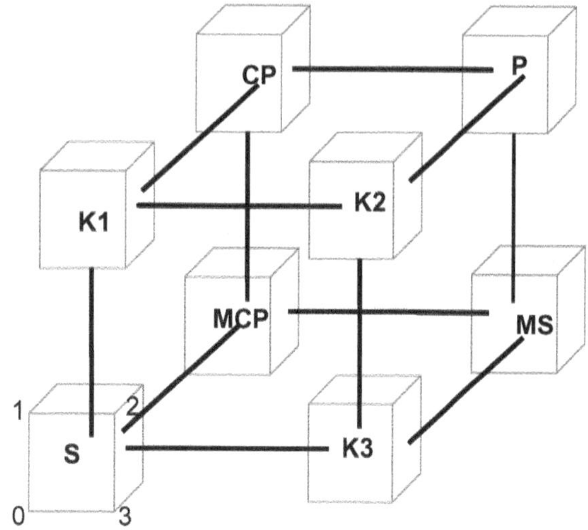

Fig. 2.16 Exploratory post formal polytope 6D

Relational complexity metric is described in terms of arguments, ai, and relations R, and can be represented as: R(ai,aj,….).

Binary relations include relations between two arguments; ternary relations involve three arguments; quaternary relations involve four interrelated arguments, and so forth.

Research assessed child capacity to perform different types of representational tasks (for example transitivity, class inclusion, hierarchical classification, cardinality, and so on) that vary in relational complexity. Relational complexity was directly related to the age of acquisition of target skills.

Relational complexity theory is not a development stage theory in the traditional sense of Piaget's theory. Instead, research is able to track gradual developmental changes in children's thinking by identifying the relational complexity of the tasks that they are able to perform.

Rank is the number of entities that are bound into a representation, and is related to the number of dimensions, which is a measure of complexity. Each rank corresponds to a class of neural nets.

Halford et al. [19] proposed that models based on processing relations capture the structure sensitivity of higher cognitive-intelligent processes while they can also be compared with more basic processes such as associations. Relations have properties that are not shared by associations.

Computational cost increases exponentially with representational rank. Thus the model provides a natural explanation for processing capacity limitations for humans. Each rank corresponds to a class of cognitive processes, neural nets, empirical criteria and dimension.

Figure 2.17 shows representational ranks and schematic neural net

H denotes the hidden layer in the net. R denotes the relation, ai denotes the arguments. The rank varies from 0 to 6.

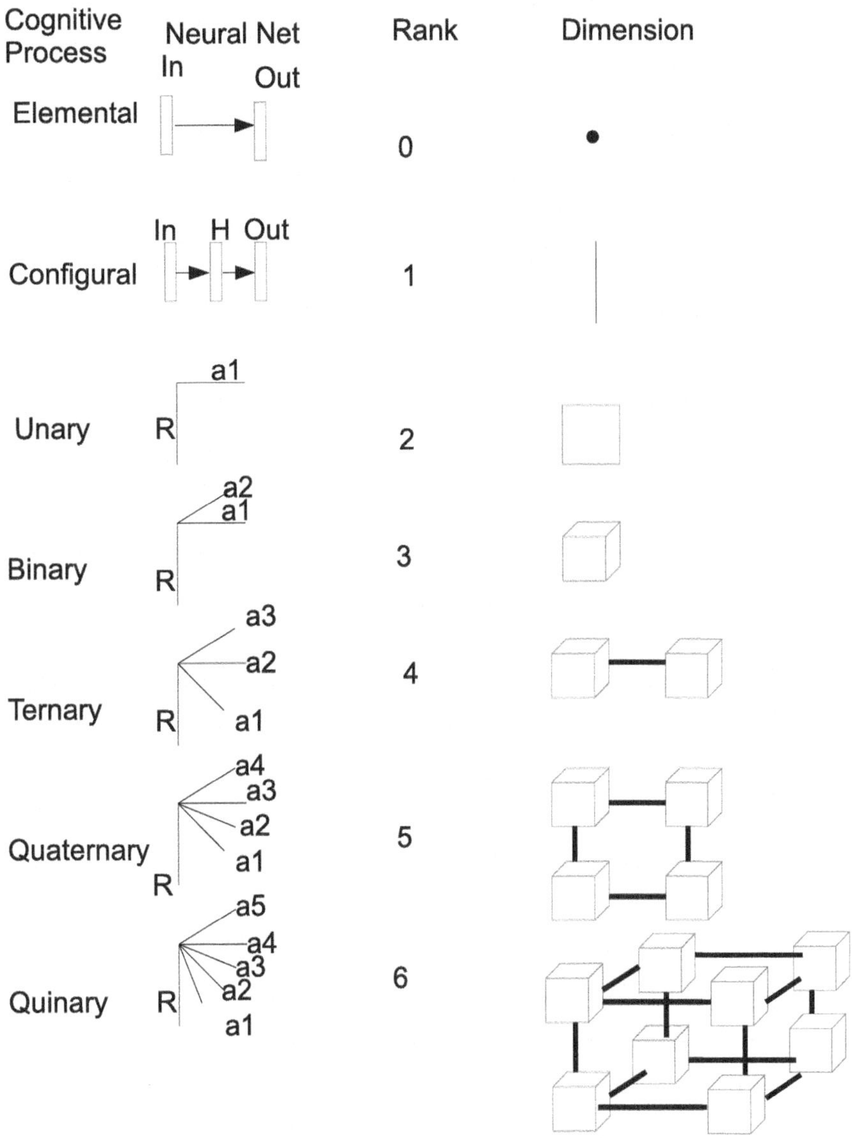

Fig. 2.17 Polytopic Roadmaps: ranks and schematic neural net

The ranks and typical concepts which belong to them are:

- Rank 0-elemental association
- Rank 1-content-specific representations and configural associations
- Rank 2-unary relations, class membership, variable-constant bindings
- Rank 3-binary relations, proportional analogies

- Rank 4-ternary relations, transitivity and hierarchical classification
- Rank 5-quaternary relations, proportion and the balance scale
- Rank 6-quinary relations.

Rank 0 can be performed by 2-layered nets, rank 1 by 3-layered nets, and ranks 2–6 by tensor products of the corresponding number of vectors.

Implications of this model have been developed for human reasoning and cognitive-intelligent development. Polytopic Roadmaps that takes into account the dimension are included in Fig. 2.17.

References

1. Baas, N.A.: Hyperstructures, topology and datasets. Axiomathes **19**, 281–295 (2009)
2. Baas, N.A.: Higher order architecture of collections of objects. Int. J. Gen. Syst. **44**(1), 55–75 (2015)
3. Baas, N.A.: On higher structures. Int. J. Gen. Syst. **45**(6), 747–762 (2016)
4. Baas, N.A., Ehresmann, A.C., Vanbremeersch, J.P.: Hyperstructures and memory evolutive systems. Int. J. Gen. Syst. **33**(5), 553–568 (2004)
5. Baheti, R., Gill, H.: Cyber-physical systems. The impact of control technology, **12**, 161–166 (2011)
6. Barnsley, M.F.: Fractals Everywhere. Academic, New York (1993)
7. Bringsjord, S., Taylor, J., Wojtowicz, R., Arkoudas, K., van Heuvlen, B.: Piagetian roboethics via category theory: moving beyond mere formal operations to engineer robots whose decisions are guaranteed to be ethically correct. In: Anderson, M., Anderson, S. (eds.) Machine Ethics. Cambridge University Press, Cambridge (2010)
8. Case, R.: The Mind's Staircase: Exploring the Conceptual Underpinnings of Children's Thought and Knowledge. Hillsdale, NJ, Erlbaum (1992)
9. Commons, M.L.: Introduction to the model of hierarchical complexity and its relationship to postformal action. World Futures **64**(5–7), 305–320 (2008)
10. Dawson, T., Heikkinen, K.: Identifying within-level differences in leadership decision making. Integr. Leadersh. Rev. **9**(5), 1–13 (2009)
11. Deleuze, G.: The Fold: Leibniz and the Baroque. University of Minnesota Press, Minneapolis (1993)
12. Ehresmann, A.C., Vanbremeersch, J.-P.: Memory Evolutive Systems: Hierarchy Emergence, Cognition. Elsevier, Amsterdam (2007)
13. Eigen, M., Schuster, P.: The Hypercycle. Springer, Berlin (1979)
14. Fischer, K.W.: A theory of cognitive development: the control and construction of hierarchies of skills. Psychol. Rev. **87**, 477–531 (1980)
15. Fischer, K.W.: Dynamic cycles of cognitive and brain development: measuring growth in mind, brain, and education. In: The Educated Brain: Essays in Neuroeducation, pp. 127–150 (2008)
16. Fischer, K.W., Bidell, T.R.: Dynamic development of action and thought. In: Damon, W., Lerner, R.M (Eds.), Volume 1. Theoretical Models of Human Development. Handbook of Child Psychology, 6th edn., pp. 313–399. Wiley, New York (2006)
17. Fischer, K.W., Rose, S.P.: Dynamic growth cycles of brain and cognitive development .In: Thatcher, R.W., Lyon, G.R., Rumsey, J., Krasnegor, N. (eds.) Developmental Neuroimaging: Mapping the Development of Brain and Behavior, pp. 263–283. Academic, San Diego (1996)
18. Halford, G.S., Andrews, G.: Information-processing models of cognitive development. Wiley-Blackwell Handbook Childhood Cogn. Dev. **16**, 697–722 (2010)
19. Halford, G.S., Wilson, W.H., Phillips, S.: Processing capacity defined by relational complexity. Implications for comparative, developmental and cognitive psychology. Behav. Brain Sci. **21**(6), 803–831 (1998)

20. Iordache, O.: Evolvable Designs of Experiments. Applications for Circuits. Wiley-VCH, Weinheim, Germany (2009)
21. Iordache, O.: Polystochastic Models for Complexity. Springer, Berlin, Heidelberg (2010)
22. Iordache, O.: Implementing Polytope Projects for Smart Systems. Springer, Cham, Switzerland (2017)
23. Iordache, O.: Advanced Polytopic Projects. Springer, Cham, Switzerland (2018)
24. Iosifescu, M., Grigorescu, S.: Dependence with Complete Connections and Applications. Cambridge University Press, Cambridge (1990)
25. Jones, D., Snider, C., Nassehi, A., Yon, J., Hicks, B.: Characterising the digital twin: a systematic literature review. CIRP J. Manuf. Sci. Technol. (2020)
26. Kelso, J.A.S., Engstrom, D.A.: The Complementary Nature. MIT Press, Cambridge, MA (2006)
27. Latour, B.: We Have Never Been Modern. Harvard University Press, Cambridge, MA (1993)
28. Lee, E.A.: The past, present and future of cyber-physical systems: a focus on models. Sensors **15**(3), 4837–4869 (2015)
29. Lehn, J.-M.: Supramolecular Chemistry: Concepts and Perspectives. Wiley-VCH, Weinheim (1995)
30. Lupasco, S.: Logique et Contradiction. Presses Universitaires de France, Paris (1947)
31. May, J.P.: The Geometry of Iterated Loop Spaces, vol. 271. Springer (2006)
32. Maier, M.W.: Architecting principles for system of systems. Syst. Eng. J. Int. Council Syst. Eng. **1**(4), 267–284 (1998)
33. Nicolescu, B.: Manifesto of Transdisciplinarity. SUNY, Albany, USA (2002)
34. Nicolescu, B.: Methodology of transdisciplinarity: levels of reality, logic of the included middle and complexity. Transdisciplinary J. Eng. Sci. **1**, 19–38 (2010)
35. Overton, W.F.: Understanding, explanation, and reductionism: finding a cure for Cartesian anxiety. In: Reductionism and the Development of Knowledge, pp. 39–62. Psychology Press (2003)
36. Overton, W.F.: Developmental psychology: philosophy, concepts, methodology. In: Lerner R.M. (ed.) Volume 1. Theoretical Models of Human Development. Handbook of Child Psychology, 6th edn, pp. 18–88. Wiley, Hoboken (2006)
37. Overton, W.F.: Relationism and relational developmental systems: a paradigm for developmental science in the post-Cartesian era. Adv. Child Dev. Behav. **44**, 21–64 (2013)
38. Piaget, J.: Genetic Epistemology. Columbia University Press, New York (1970)
39. Piaget, J.: The Construction of Reality in the Child. Ballantine Books, New York (1971)
40. Piaget, J.: Introduction a l'epistemologie genetique. Presses Universitaires de France, La pensee Mathematique, Paris, France (1973)
41. Piaget, J., Garcia, R.: Psychogenesis and the History of Science. Columbia University Press, New York (1989)
42. Tao, F., Qi, Q., Wang, L., Nee, A.Y.C.: Digital twins and cyber–physical systems toward smart manufacturing and industry 4.0: correlation and comparison. Engineering **5**(4), 653–661 (2019)
43. von Bertalanffy, L.: General Systems Theory: Foundations, Development, Applications. George Braziller, New York (1968)
44. von Uexküll, J.: Theoretische Biologie. Suhrkamp Taschenbuch Wissenschaft, Frankfurt a. M. (1973)
45. Wiener, N.: Cybernetics or Control and Communication in the Animal and the Machine. MIT Press (2019)
46. Yaghi, O.M.: Reticular chemistry in all dimensions. ACS Central Science **5**(8), 1295–1300 (2019)

Chapter 3
Operations

3.1 Partitions

3.1.1 Integers Partition

The partitions have many applications in science and engineering fields. The separations and the chemical decompositions can be studied as partitions.

Partitions can be used to model the possible outcomes of fission, where the resulting fragments of the nucleus correspond to a partition of the total number of components in the samples.

Mathematical concepts closely related to partitions have applications in physics such as Young tableaux, which are essential in the analysis of the symmetries of many physical systems [7, 39]

Applications for partitions also arise in biochemistry. In population genetics partitions are used to model the genetic variation of a set of gametes from a large population [17].

Partitions have connections with number theory, representation theory, derivatives and symmetric groups polynomials.

Combinatorial enumeration problems arise in many problems of statistics [24]. The number of partitions of an integer is one such enumeration problem with a history dating back to Euler [29]. Examples of applications to statistical physics and material science problems include the Potts model, crystal growth, lattice polygons, Bose Einstein statistics, and polymer coverings.

The solution to the integer partitioning problem is known for 1-dimensional and 2-dimensional partitions.

Consider for example the partition of the integer 4 shown in Fig. 3.1.

Figure 3.1a shows the 1-dimension partition while Fig. 3.1b shows the 2-dimension partition.

Not much is known about higher dimensional partitions. Numerical estimation of the asymptotic behavior of these higher dimensional partitions could lead to new

O. Iordache, *Polytopic Roadmaps*, Studies in Systems, Decision and Control 368 ,
https://doi.org/10.1007/978-3-030-75630-7_3

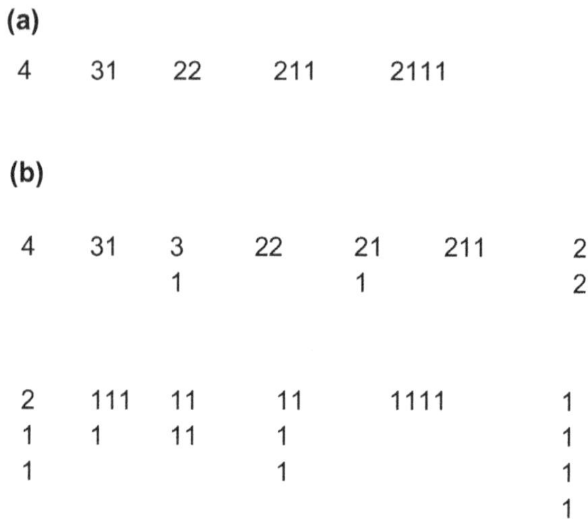

Fig. 3.1 Partition of the integer 4

theoretical insights. The 1-dimensional or linear partition of an integer is the decomposition into a sum of positive integers in which the summands are ordered from largest to smallest.

The 2-dimensional or plane partition of an integer is the decomposition into a sum of smaller positive integers which are arranged on a plane. The ordering property generalizes to the summands being un-increasing along both the rows and the columns.

The 3-dimensional or spatial partition of an integer is the decomposition into a sum of smaller positive integers which are arranged in space.

Generalization to d-dimension is straightforward.

Consider a d-dimensional hyper cubic lattice with sites labeled by $i = (i_1, i_2,..., i_d)$, where $i_k = 1, 2,....$ An integer height h(i) (corresponding to a summand) is associated with site i.

A d-dimensional partition of a positive integer n is a configuration of heights such that:

$h(i) \geq 0$, $h(i) \geq \max_{1 \leq k \leq d} h(i_1, i_2,..., i_k + 1,..., i_d)$, $\sum_i h(i) = n$. This means that the heights h(i) are un-increasing in each of the d-lattice directions. Figure 3.1 illustrates this condition for linear and plane partitions.

3.1.2 *Dual Graded Graphs for Integers Partitions*

A partition of a positive integer n is an unordered collection of positive integers whose sum is n. For example the three partitions of the integer 3 are: 3, 2 + 1 and 1 + 1 + 1.

Since we refer to sum there is no difference between 2 + 1 and 1 + 2.

Figure 3.2 illustrate the partition of an integer.

A differential poset is a partially ordered set, or poset for short, satisfying certain local properties [27, 28]. These posets have been introduced by Stanley as a generalization of Young's lattice (the poset of integer partitions ordered by inclusion), many of whose combinatorial properties are shared by all differential posets.

Young diagrams have been associated to integer partitions. They illustrate facts as for instance, that there are three partitions of 3: (3), (2, 1) and (1, 1, 1).

The concept of differential poset was generalized in different ways.

Significant studies of dual graded graphs, (DGG), are due to Fomin [5, 6].

Fomin replaced the differential poset by two graphs. The two graphs in duality are typically referred as U and D graphs. They have the same set of vertices.

U and D correspond to the up and down directions for the graphs.

U corresponds to constructions or combinations, D to deconstructions or decompositions.

The procedure developed by Fomin can be considered as a basic process of doubling and relating. This is one of the main procedures that generate Polytopic Roadmaps.

The relating after doubling is presented by Fomin in terms of commutation relations.

The D graphs play a role analogous to that of the digital part or the cyber part in cyber-physical systems CPS or in digital twins DT [25, 32]. The U graph is more close to the physical part of CPS or the physical model for digital twins.

CPS is in the core of industry 4.0 developments and DT for modeling. Consequently DGG are important for industrial developments.

Figure 3.3 shows the U-graph for integer partition [9].

Figure 3.4 shows the D-graph for integer partition [9].

Observe that U and D have the same elements but the interconnections between elements make the difference. The unique minimal element is denoted here by $\hat{0}$.

The weights indicated on the connections between two elements in the graphs show the number of ways we can obtain one from another. This depends on the selected insertion rules.

Significant for DGG is the rule connecting the two graphs in duality.

Figure 3.5 illustrates the commutation condition for integer partition

The commutation condition compares the evolution on the dual graded graphs in the order U then D, to the order D then U. The result shows that, for tree-like graphs as shown in Fig. 3.3 it is preferable a steps in the U then D directions since the U step involves more nodes than the D step and allows more opportunities for evolution.

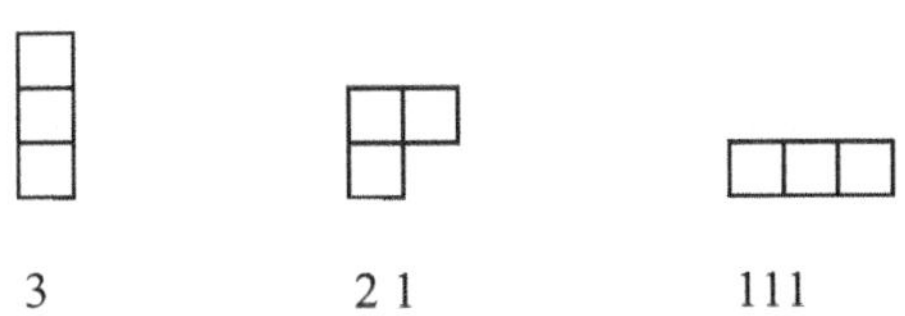

Fig. 3.2 Illustrative partition of integers

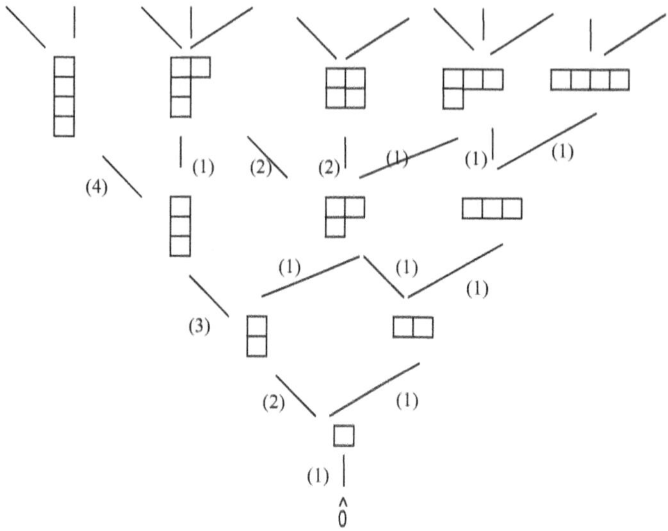

Fig. 3.3 U-graph for integer partition

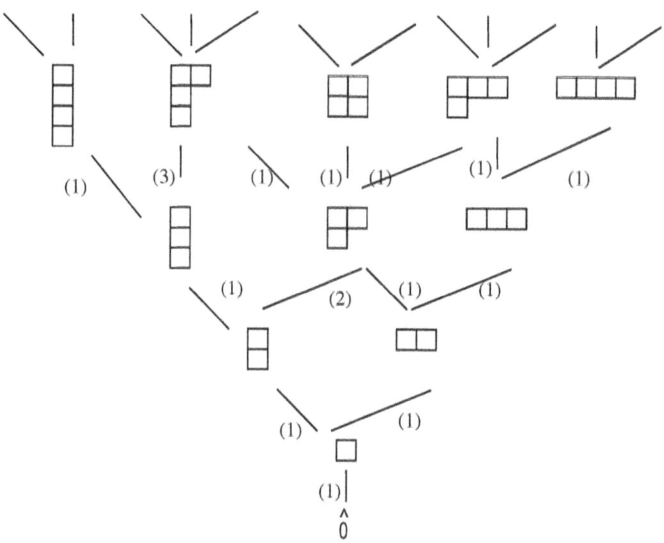

Fig. 3.4 D-graph for integer partition

A method to increase dimensionality starting from dual graded graphs is illustrated in Fig. 3.6. This should be considered as a continuation and generalization of the DGG procedure.

Figure 3.6 shows the Polytopic Roadmaps associated to partitions.

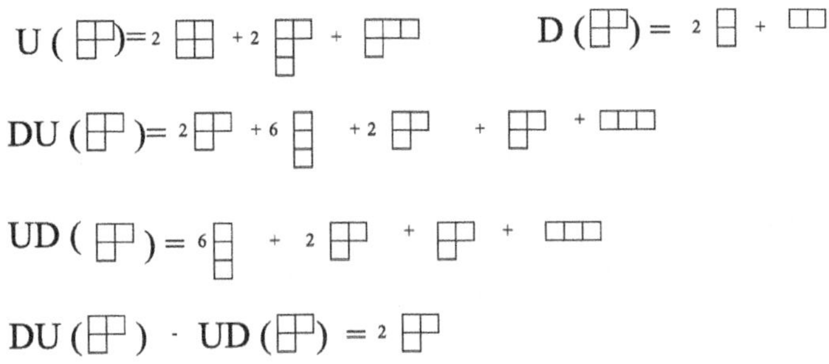

Fig. 3.5 Commutation condition for integer partition

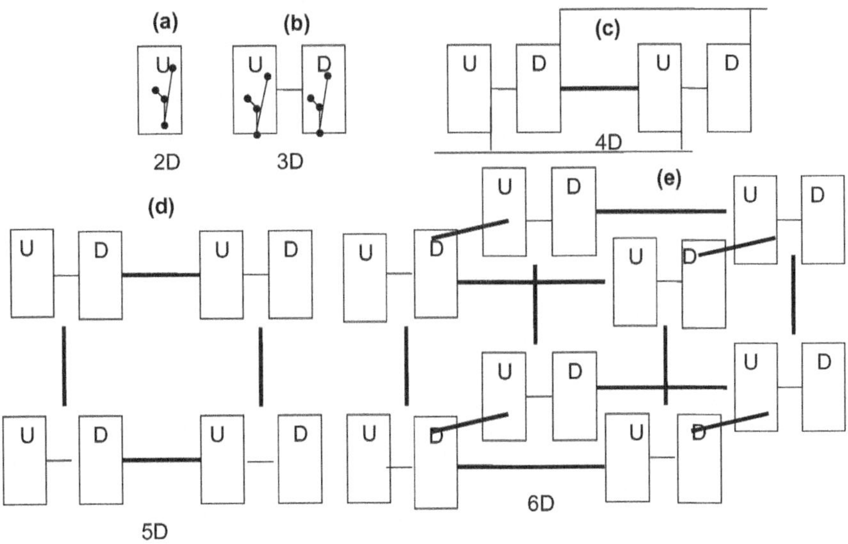

Fig. 3.6 Polytopic Roadmaps for integer partitions

The starting diagram is a U-graph (2D) or a pair U, D-graphs as shown in Fig. 3.3 and Fig. 3.4. The pair U, D results from U by doubling or twinning followed by relating. It is a 3D polytope, the U-D cube as shown in Fig. 3.6b. The U and D trees may be considered in different planes and this justifies why U-D is associated to a cube. The relating is represented in Fig. 3.6b by a thin line. The doubling is not limited to one step as for DGG procedure but may be continued.

The next doubling or twining increase dimensions from 3 to 4D. A new relation is shown by a thick line in Fig. 3.6c. The previous relation between U and D is maintained but new relations should be considered between the two cubes U-D. A proposal in which only U graphs or D graphs are involved in relating is shown in Fig. 3.6c.

Continuing the doubling and relating we obtain the 5D frame from Fig. 3.6d and then the 6D frame from Fig. 3.6e. The relating is specific for different dimensions and this may ensures the system versatility and potential for self-evolvability.

It should be noted that the Polytopic Roadmaps shown in Fig. 3.6 are general and may be applied to other pairs of dual graded graphs U, D.

Figure 3.7 shows a Periodic Table for integer partitions.

A linear that is 1D partition completes the framework. This corresponds to a partition in which each cell contains a 1 element of the integer. For instance, 4 is represented as 1111.

The first row of the Periodic Table shows the elements, 1D, 2D and 3D.

The second row contains the elements 4D, 5D, and 6D.

The first column corresponds to the line (1D), the second to the square or rectangle (2D) frame, while the third to a cube (3D). Second row outline the same sequences, line, square, cube as patterns or skeletons.

Figure 3.8 shows a Periodic Table for any pair of dual graded graphs, U and D.

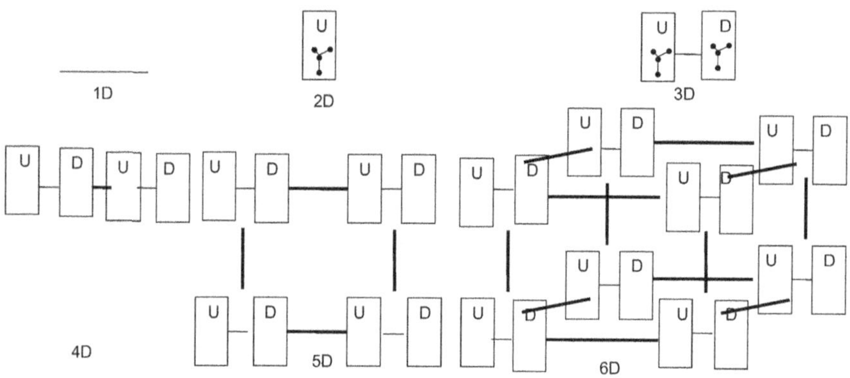

Fig. 3.7 Periodic Table for integer partitions

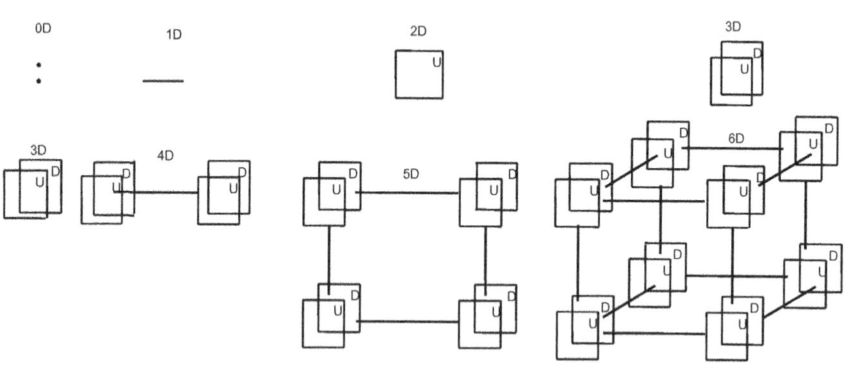

Fig. 3.8 Periodic Table for dual graded graphs

The 0D and 1D component completes the previous frameworks.

The 0D corresponds to un-related components.

The first row of the table shows the elements 0D, 1D, 2D and 3D.

The second row contains the elements 3D, 4D, 5D, and 6D.

The first column corresponds to the dots (0D) the second to the line (1D), the third to the square (2D) frame, while the fourth to a cube (3D). Second row outline these patterns or skeletons.

3.1.3 K-Bounded Partitions

For a k-bounded partition the maximum height is k. Such partitions describe evolution in restricted domains.

For 4 we have the 2-partition: 22, 211, and 1111. The maximum height is 2.

The U tree and the D tree are presented in Fig. 3.9.

U and D correspond to the up and down trees or graphs.

The weights indicated on the connections between two partitions show the number of ways we can obtain one from another. The weights are different in the U-graph and D-graph.

Figure 3.10 shows an example of commutation condition.

Polytopic Roadmaps and Periodic Tables may be created for k-bounded partitions.

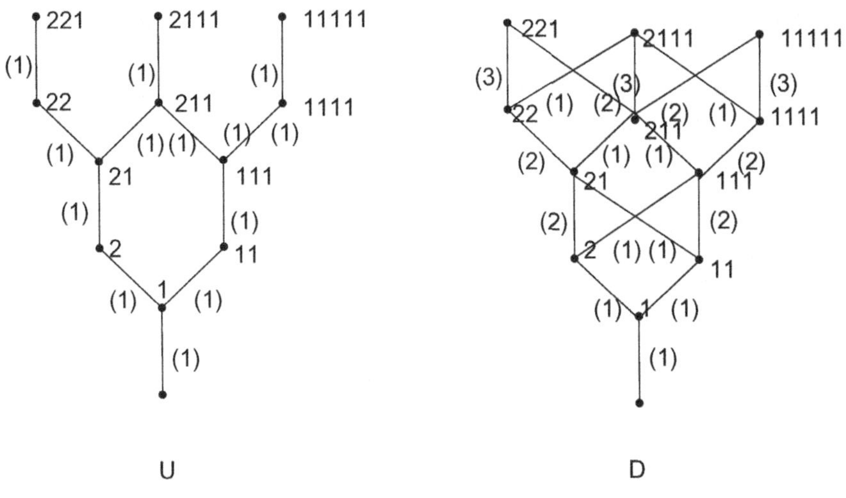

Fig. 3.9 U and D graphs for k-Young lattice, k = 2

Fig. 3.10 Commutation condition for k-Young lattice

U(2)=21 D(2)=1

DU(2)=2x2+11 UD(2)=2+11

(DU-UD)(2)=2

3.2 Permutations

3.2.1 Permutation Trees for Designs of Experiments

Simple designs of experiments (DoE) matrices may be associated to different permutations.

Figure 3.11 shows equivalent notations for permutations

We associate the permutation 231 to the DoE matrix in which for any experiment just one factor is modified. Experiments are assigned to rows and factors to columns in that matrix.

For the hierarchy of experiments and factors we may use the permutation trees [6]. Figure 3.12 shows the dual graded graphs for permutation trees.

Figure 3.12a shows the U graph while Fig. 3.12b shows the D graph.

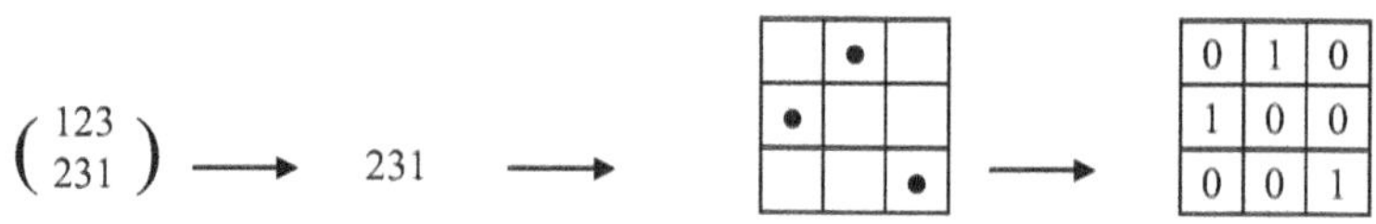

Fig. 3.11 Equivalent notations for permutations

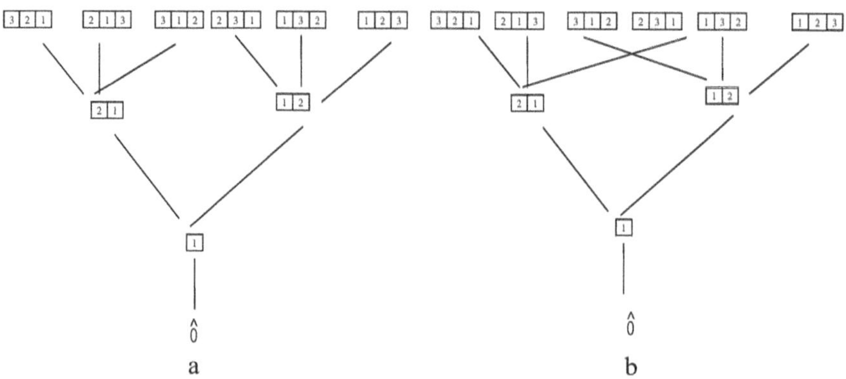

Fig. 3.12 Dual permutation trees

$$U(21) = 321 + 213 + 312 \qquad D(21) = 1$$

$$DU(21) = 21 + 21 + 12$$

$$UD(21) = 21 + 12$$

$$DU(21) - UD(21) = 21$$

Fig. 3.13 Commutation condition for permutation trees

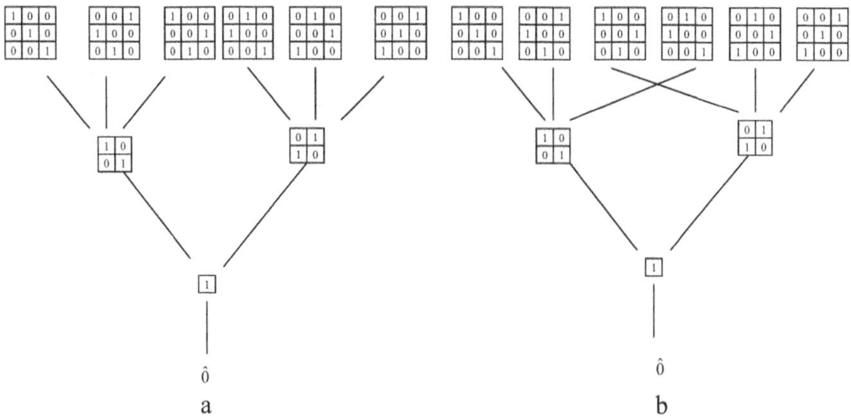

Fig. 3.14 Dual permutation trees of designs

Figure 3.13 illustrates the commutation condition for permutation trees.

The permutation graphs may be considered as trees of elementary designs of experiments. Figure 3.14 shows the dual permutation trees of designs of experiments.

Figure 3.14a shows the U graph while Fig. 3.14b shows the D graph.

The method to increase dimensionality starting from dual graded graphs is illustrated in Fig. 3.15. This figure shows Polytopic Roadmaps associated to permutations.

The starting diagram is a U-graph (2D) or a pair U, D-graphs as that shown in Fig. 3.12 or Fig. 3.14. The pair U, D results from U by doubling or twinning followed by relating. Observe that U and D contain the same nodes. It results a 3D polytope, a cube as shown in Fig. 3.15b. The relating is indicated here by a thin line U-D. The relation between U and D may be that from commutation operation, that is simple translations.

Flexible relations transform DoE in evolvable DoE, (EDoE).

The next doubling increase dimension to 4D. A new relation is shown by a thick line in Fig. 3.15c. The previous relation between U and D is maintained but new relation should be considered between the two cubes U-D. A first proposal in which only U graphs or D graphs are involved is shown in Fig. 3.15c.

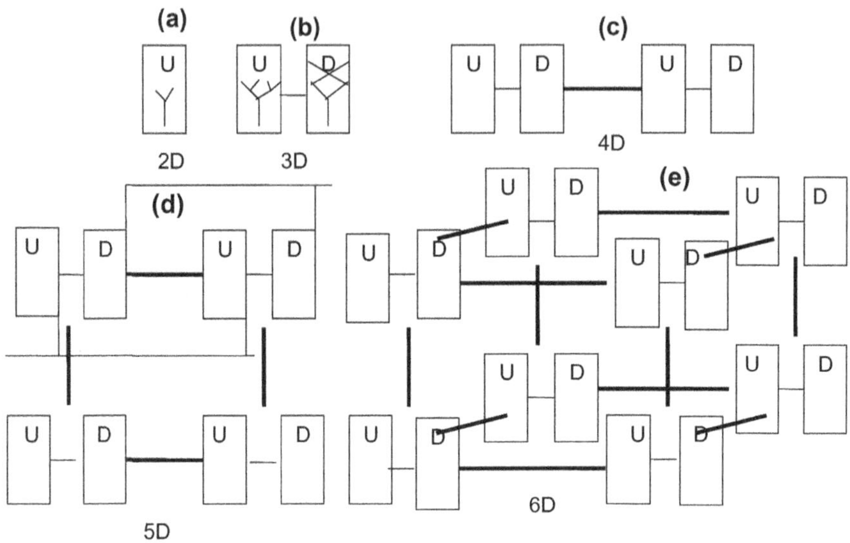

Fig. 3.15 Polytopic Roadmaps for permutations

By doubling and relating procedure we may obtain the 5D frame from Fig. 3.15d and then the 6D frame from Fig. 3.15e. The relating is different for different dimensions. This is the source of frameworks adaptability and self-evolvability for DoE and EDoE.

3.2.2 *Latin Hypercubes*

A permutation can be encoded by means of a permutation matrix. This is an $n \times n$ array of 0 and 1 in which every line contains exactly one 1-entry. Here a line means either a row or a column.

An order-n permutation is an $n \times n$ (0,1)-matrix with exactly one 1 in each row and column. An order-n two-dimensional permutation is an $n \times n \times n$ (0,1)-array with exactly one 1 in each row, column, and shaft.

This suggests the definition of a d-dimensional permutation on [n] [21]. It is an array $[n] \times [n] \times \ldots \times [n] = [n]^{d+1}$ (with $d + 1$ factors) of zeros and ones in which every line contains exactly one 1-entry. Whereas a matrix has only two kinds of lines, namely rows and columns, now there are $d + 1$ kinds of lines. A line is a set of n entries in the array that are obtained by establishing d of the $d + 1$ coordinates and letting the remaining coordinate take all values from 1 to n.

Just as permutation matrices are equivalent to permutations, 2-permutations are equivalent to Latin squares.

The definition yields a one-to-one correspondence between 3-dimensional permutations and Latin cubes. In general, d-dimensional permutations are synonymous with d-dimensional Latin hypercubes.

Consequently we have the succession of correspondences:

1D– 1-permutation (1-polystochastic)

2D – 2-permutation (Latin square),

3D – 3- permutation (Latin cube)

nD – n-permutation (nD Latin hypercube)

The d-dimensional matrix is called 1-polystochastic if it is non-negative and the sum over each line equals 1. Such a matrix that has a single 1 in each line and zeros elsewhere is called a 1-permutation matrix. Notice that the term polystochastic was used with a different meaning to describe processes of processes and so on in hierarchical conditioning [11].

The relation between DoE and Latin squares is significant for industrial processes since to any DoE a process design may be associated. DoE offers the processing parameters.

Figure 3.16 shows a Latin cube of order 4 [23]

The three dimensions of the Latin cube are referred to as columns, x_1, rows, x_2, and pages or sheets. The cube shown in Fig. 3.16 contains four pages. A cube and a developed form have been presented in Fig. 3.16.

It can be verified from Fig. 3.16 that in each one of the four pages, no two entries repeat in a row as well as a column. Similarly, for a fixed value of column index, the entries in the four pages are distinct.

Figure 3.17 shows a Latin hypercube of dimension 4 and order 4.

In Fig. 3.17 with a 4 × 4 square for which the indices of the third and fourth dimensions are fixed, no two entries repeat in a row as well as a column. When the first three dimensions are fixed, no two entries repeat when the fourth dimension is varied. Fixing the first, second and fourth dimensions, the four entries obtained when the third dimension is varied are distinct.

Applications of Latin designs to describe cycling separations scheduling have been presented by [11, 12].

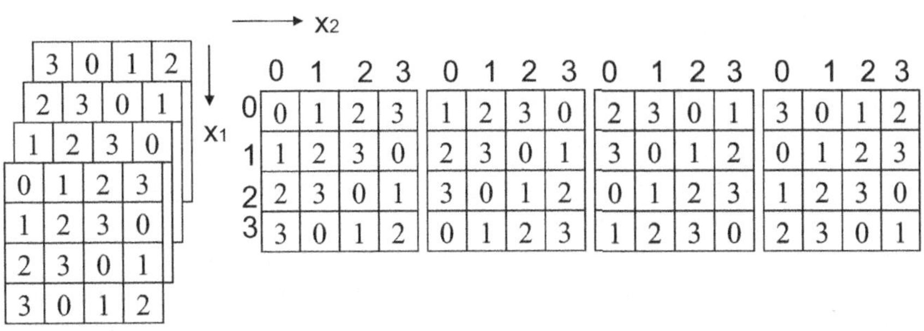

Fig. 3.16 Latin cube of order 4

		$x_3 = 0$				$x_3 = 1$				$x_3 = 2$				$x_3 = 3$			
	x_1 \ x_2	0	1	2	3	0	1	2	3	0	1	2	3	0	1	2	3
$x_4=0$	0	3	0	1	2	0	1	2	3	1	2	3	0	2	3	0	1
	1	2	3	0	1	3	0	1	2	0	1	2	3	1	2	3	0
	2	1	2	3	0	2	3	0	1	3	0	1	2	0	1	2	3
	3	0	1	2	3	1	2	3	0	2	3	0	1	3	0	1	2
$x_4=1$	0	1	2	3	0	2	3	0	1	3	0	1	2	0	1	2	3
	1	0	1	2	3	1	2	3	0	2	3	0	1	3	0	1	2
	2	3	0	1	2	0	1	2	3	1	2	3	0	2	3	0	1
	3	2	3	0	1	3	0	1	2	0	1	2	3	1	2	3	0
$x_4=2$	0	2	3	0	1	3	0	1	2	0	1	2	3	1	2	3	0
	1	1	2	3	0	2	3	0	1	3	0	1	2	0	1	2	3
	2	0	1	2	3	1	2	3	0	2	3	0	1	3	0	1	2
	3	3	0	1	2	0	1	2	3	1	2	3	0	2	3	0	1
$x_4=3$	0	0	1	2	3	1	2	3	0	2	3	0	1	3	0	1	2
	1	3	0	1	2	0	1	2	3	1	2	3	0	2	3	0	1
	2	2	3	0	1	3	0	1	2	0	1	2	3	1	2	3	0
	3	1	2	3	0	2	3	0	1	3	0	1	2	0	1	2	3

Fig. 3.17 Latin hypercube of dimension 4 and order 4

3.3 Classifications

3.3.1 Conceptual Lattices for Classification

Classification methods have been applied in chemical engineering for separation flow-sheets generation. Similarity based clustering method and informational criteria allow generating flow-sheets on the bases of difference in properties responsible for separation [14, 15]. For example difference in absorbability allows separation in absorbers that is absorption columns, while difference in volatility allows separation in distillation columns.

Formal Concepts Analysis (FCA) methods can be applied for such classification problems. The result will be conceptual separation schemes.

FCA highlights the chemical compound separation schemes and also the properties correlated to the separation devices.

Studies concerning FCA and classifications are numerous. A comparison between similarity-based clustering and classification lattices methods is done by Valtchev and Missaoui [35]. Clearly the separation scheme depends on classification procedure.

Interesting adaptations of FCA that allow the systematic analysis of drugs structure–activity and structure-selectivity relationship have been proposed by Lounkine and Bajorath [22]. Fragment FCA assesses the distribution of molecular fragment combinations among ligands with closely related biological targets. Molecular FCA was introduced for the systematic comparison of the selectivity of a compound against multiple targets and the extraction of compounds with complex selectivity profiles from biologically annotated databases.

The starting point for FCA application for classification and separation is a context table as Table 3.1 that contain objects or compounds to be separated and properties or attributes allowing to separate.

To such tables a conceptual lattice may be associated Ganter and Wille [8].

Table 3.1 shows an example of formal context with the set of objects G = {1, 2,...,6} and the set of properties M = {a, b, c, d}. In this case the objects are chemical compounds [14, 15]. The properties show high values denoted by"1″ or low values denoted by "0″. To any property, specific separation equipments may be associated.

Figure 3.18 shows the conceptual lattice for classification based on four properties.

The conceptual lattice is associated to data presented in Table 3.1.

At the top is the union of all components. The null symbol Φ indicates that no property was activated to operate classifications.

Each marked points in conceptual lattice is labeled with both the components or objects and the properties that define the separation.

Above and to the right of the diagram we can see that the compound 4 resulted in the light phase from the separating devices corresponding to b, d, and c as properties.

To the left of the diagram we see that the compound 1 was separated by the separating devices corresponding to a, d and c as properties.

Table 3.1 Formal context for classifications-four properties

Compounds	a	b	c	d
1	1	0	1	1
2	1	0	0	1
3	1	0	0	0
4	0	1	1	1
5	0	1	0	0
6	0	1	0	1

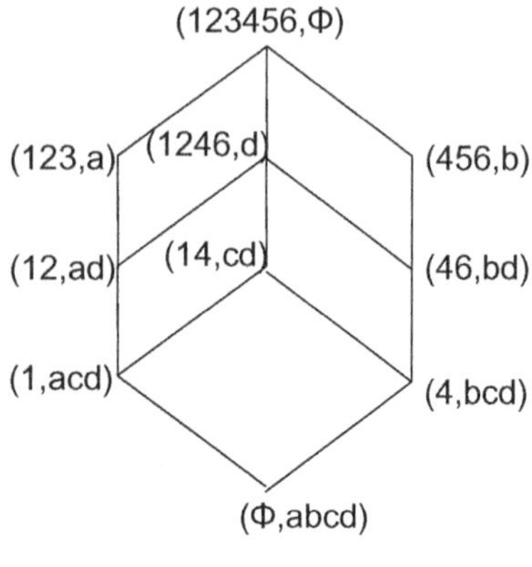

Fig. 3.18 Concept lattice for classification-four properties

As we move down, on a trajectory, we encounter smaller collections of compounds and larger collection of properties that is of separating equipments in the separation scheme or flow-sheet.

At the bottom of the diagram we see that there are no compounds having all the properties.

3.3.2 Polytopic Formal Concept Analysis

Designing efficient data mining algorithms to compute collections of relevant patterns is an active research domain. Useful knowledge discovery processes can be based on patterns extracted from large multi-dimensional datasets.

Many datasets record whether some properties hold for some objects, for instance, whether a product is bought by a customer or whether a gene is over-expressed in a biological sample. Such datasets are binary relations and can be represented as (0, 1) matrices. In such matrices, a closed item set is a maximal rectangle of "1"s modulo arbitrary permutations of the lines, that is objects, and the columns that is, properties. Thus, every closed item set supports the discovery of a maximal subset of objects sharing the same maximal subset of properties.

Efficiently extracting every closed item set satisfying user-defined relevancy constraints have been extensively studied [2].

This framework sometimes turns out to be less effective, since many datasets are n-ary relations, that is, (0, 1) tensors. Reducing their analysis to two dimensions is ignoring potentially interesting additional dimensions, for instance, when a gene expression is measured. The presence of noise in most real-life datasets is a second issue for low effectiveness, which leads to the fragmentation of the patterns to discover.

Conventional formal context consists of two sets, that of objects and that of attributes, together with a binary relation between objects and attributes. This relation induces in the standard way the so-called Galois connection between sets of objects and sets of attributes whose closed sets, the formal concepts, form a complete lattice, the lattice of formal concepts [8]. Because it employs a binary relation, a formal context is a 2D frame.

The triadic concept analysis was introduced as a generalization of formal concept analysis to three dimensions. This is a 3D frame for FCA. Apart from the sets of objects and attributes, a third set, called the set of conditions, was introduced and a ternary relation between objects, attributes and conditions took the place of the binary relation of formal contexts [4, 18, 40].

Voutsadakis introduced the polyadic concept analysis, which generalizes triadic contexts and concepts to n dimensions for arbitrary n [36]. This is an nD frame.

A different approach to high dimensional FCA based on the Polytopic Roadmaps is presented in what follows.

Figure 3.19 illustrates a Polytopic Roadmaps for classification.

Here o1, o2 denotes the objects, a, b, c denotes the attributes or properties.

Figure 3.19a shows the objects.

Figure 3.19b shows the separation based on a single property.

Figure 3.19c shows the separation based on several properties. It corresponds to a typical context table. It is 2D frame. The two dimensions are the objects and the attributes.

A new dimension A, B, C is shown in Fig. 3.19d. It results a 3D cube.

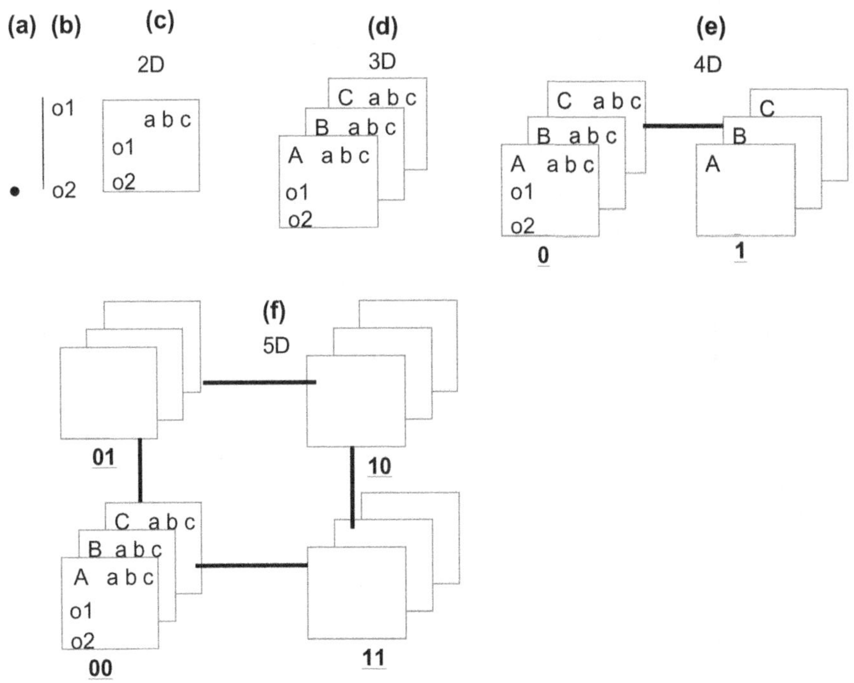

Fig. 3.19 Polytopic Roadmaps for classification

The 3rd dimension may correspond, for instance, to the installations A, B or C, selected for separation.

A, B or C represents conditions.

A new dimension indexed by **0** and **1** is shown in Fig. 3.19e.

This may corresponds to two different plants indexed by **0** and **1**. It is a meta-condition that is a condition of conditions.

A new dimension indexed by supplementary **0** and **1** is shown in Fig. 3.19f.

This may correspond to the existence of two different companies, **0** and **1,** each having two plants, **0** and **1**. The company indexed by **0** has two plants indexed by **00** and **01**.

The company indexed by **1** has two plants indexed by **10** and **11.**

A new dimension indexed by supplementary **0** and **1** is shown in Fig. 3.20. It corresponds to a meta-meta-condition.

This new level of conditioning may correspond to the market for instance to West market **0** and to East market **1**.

Figure 3.21 shows a Periodic Table for classifications.

The columns correspond in succession to just one property or attribute (lines), to objects-properties tables (rectangles), to several objects-properties tables (cubes), that are sets of parallel classification schemes for the same objects and properties.

The first row starts from one line and finish with a 3D-cube.

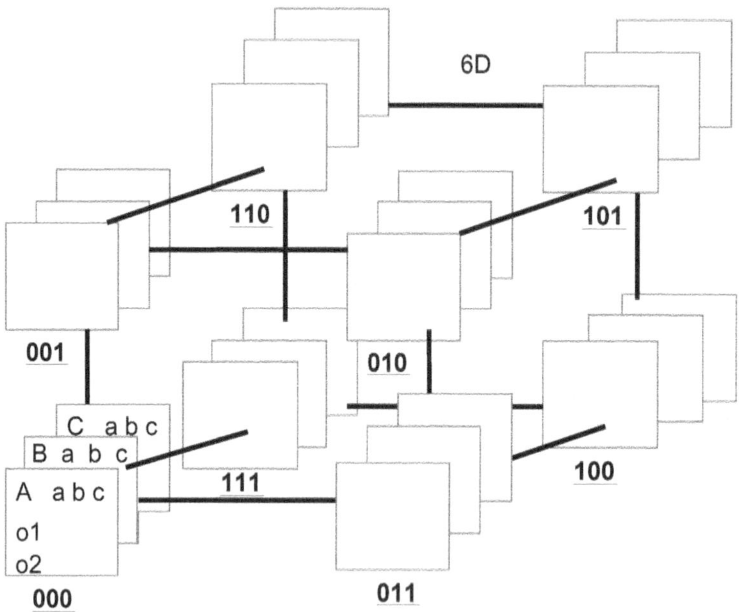

Fig. 3.20 Polytope 6D for classification

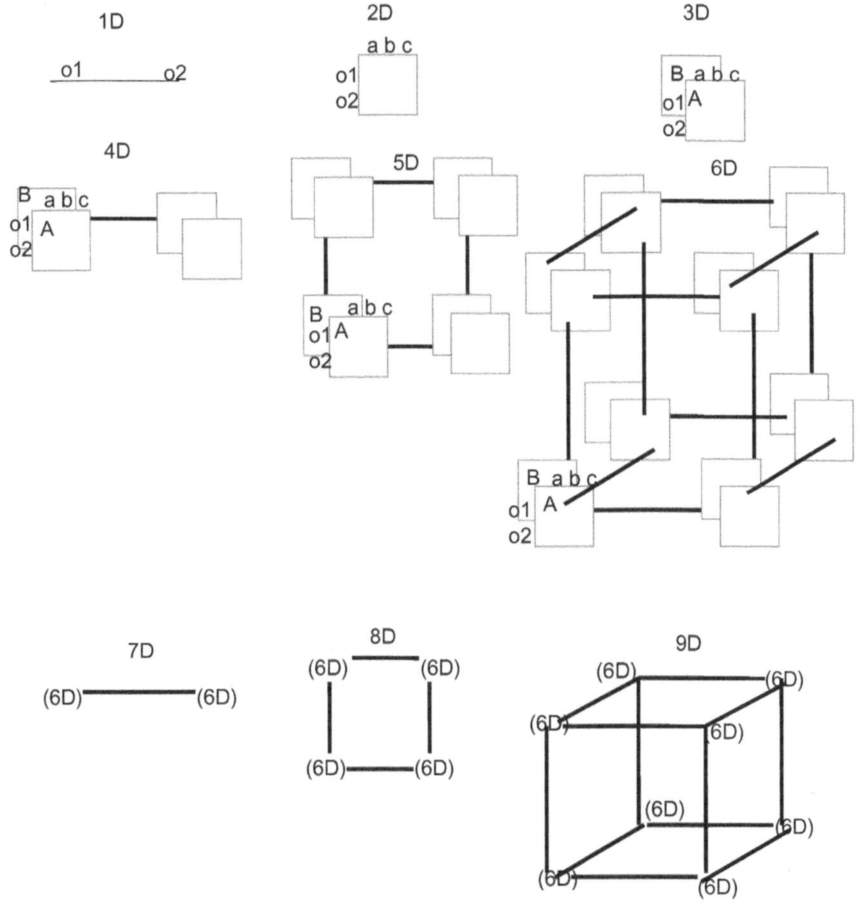

Fig. 3.21 Periodic Table for classification

The first element of the table shows two objects o1 and o2 characterized by just one value of the property, for instance high volatility. The objects can't be separated by equipments activating that property.

The second row of the Periodic Table starts from line relating two 3D-cubes (4D-cube) and finishes with cube of cubes represented as 6D-cubes.

The third row starts from line relating two 6D-cubes (7D-cube) and finish with 9D-cubes (cube of cubes of cubes).

Here, o1, o2 denotes objects to be separated on the basis of properties like a, b, and c.

A, B may denote different installations available for classification or separation. Together they are represented as a cube. The complexity of classification depends on the relations between tables and cubes as shown by thick-lines detailed presentation.

Relating is done through edges of the cubes allowing interchanging rows or columns pertaining to different objects-properties tables.

The presented high dimensional approach to FCA, keeps the context in separated sheets that allows interactions. It is an application of doubling and relating procedure for Polytopic Roadmaps.

3.3.3 Polytopic Temporal Concept Analysis

Temporal concept analysis, TCA, is an extension of FCA in which the evolutions of the objects are considered in conjunction with the conceptual aspects of the object. Significant researches in the area are due to Wolff [42–45].

Wolff has approached temporal concept analysis by scaling the time and event space and adding directed edges to the concept lattice of the context. The potential difficulty of this approach is related to the complexity of the display. Complex information bases may rapidly overcome an advantages lattice representation brings to formal concept analyses.

A way around this complexity issue is to redefine how we think about systems/objects and the states of those systems. Traditionally, we view a system in a specific state as a unique object, so we are forced in a FCA paradigm to replicate an object as many times as we have states for it. The system view as being unique with sets of constant or time dependent attributes allows reducing the complexity of the lattice.

A different approach to TCA based on the Polytopic Roadmaps and high dimensional time is presented in what follows.

Figure 3.22 illustrates this different approach to temporal concept analysis. The key is the higher dimensionality for time.

The 4D case, shown in Fig. 3.22 may be associated to the linear time as a 4th dimension.

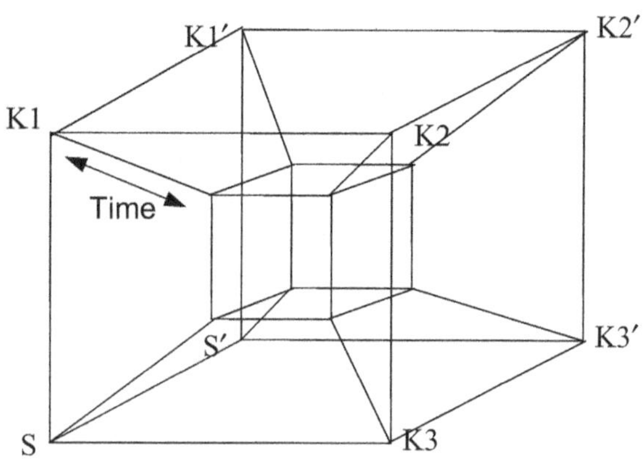

Fig. 3.22 Polytope 4D for temporal concept analysis

An example is the 4D printing that refers to printing of 3D objects that after printing, changes in time [34]. The time line connects the two 3D objects.

Figure 3.23 shows details of different presentations of 4D frame for temporal concept analysis.

It is the linear time revealed by the thick hyper-line in Fig. 3.23a, that is the line of cubes. This relates the cubes indexed by **0** and **1** in Fig. 3.23.

These **0** and **1** may correspond to different times.

These **0** and **1** correspond to external and internal cube shown in Fig. 3.22.

Recall that higher dimensions may be associated to time but also to a variation of property like, temperature, pressure, color, chemical composition and so on.

Figure 3.24 shows a Periodical Table for TCA for higher dimensions.

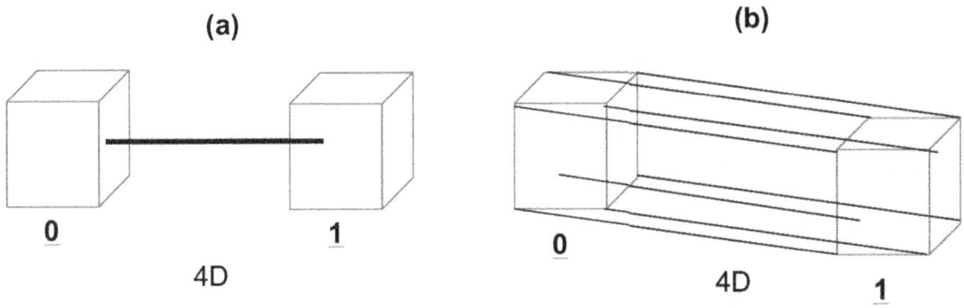

Fig. 3.23 Polytope 4D for temporal concept analysis-different presentations

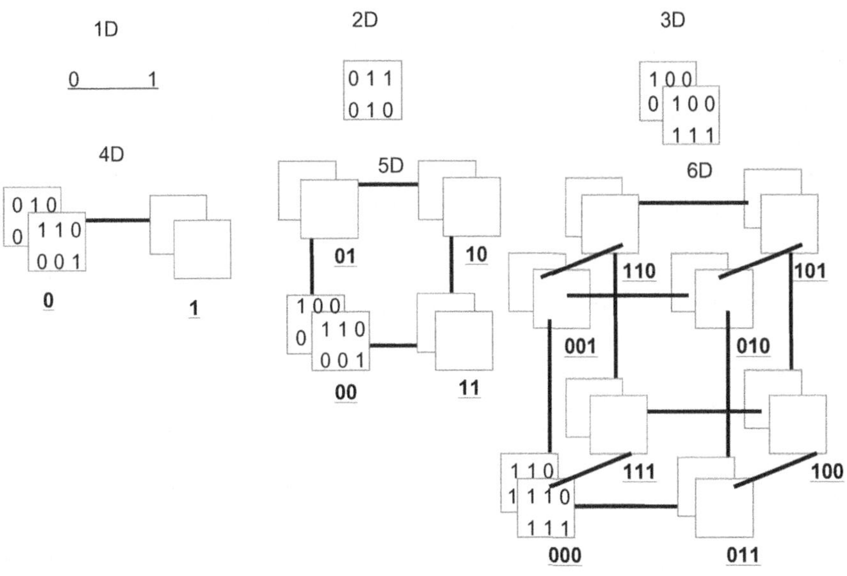

Fig. 3.24 Periodic Table for temporal concept analysis

The 4D case, shown in Fig. 3.24 is associated to the linear time as discussed.

The 5D case, will add a supplementary dimension for time. The planar time may be a possible physical interpretation for this 5D frame. It outlines a cycle of hyper-lines in plane that is a square of cubes. The 5D object refers to a 3D object that varies in planar time. Cyclical time is an example of planar time. In this case the cycle is: $\underline{\mathbf{00}} \rightarrow \underline{\mathbf{01}} \rightarrow \underline{\mathbf{10}} \rightarrow \underline{\mathbf{11}} \rightarrow \underline{\mathbf{00}}$. The transition from $\underline{\mathbf{11}}$ to $\underline{\mathbf{00}}$ closes the loop and ensures cyclicality.

The 6D situations add a supplementary dimension for time. The spatial time may be a possible physical interpretation in this case. It outlines a new cycle, and a cube of cubes resulting as a kind of parallel universes model. The front and the back faces of the 6D cube correspond to the two parallel universes. The 6D object refers to a 3D object that varies in spatial time. Helical two-cyclical time is an example of such spatial time..Multi-cyclical time allows modeling the transitions between several parallel universes.

The sequence may be: $\underline{\mathbf{000}} \rightarrow \underline{\mathbf{001}} \rightarrow \underline{\mathbf{010}} \rightarrow \underline{\mathbf{011}} \rightarrow \underline{\mathbf{100}} \rightarrow \underline{\mathbf{101}} \rightarrow \underline{\mathbf{110}} \rightarrow \underline{\mathbf{111}} \rightarrow \underline{\mathbf{000}}$.

The transition from $\underline{\mathbf{111}}$ to $\underline{\mathbf{000}}$ closes the loop and ensures cyclicality.

In fact cycles are replaced by spirals.

3.4 Complexifications

3.4.1 Graphs Spanning

Another example of differential poset is examined here [26].

Let (Em, ≤) be the set of unlabelled spanning subgraphs of Km, the complete graph on m vertices, ordered by G ≤ H if and only if G is a spanning subgraph of H.

This poset is ranked, with rank (G) equal to the number of edges of G. Under this order, the cover relations are given by G < H if and only if G is a spanning subgraph of H with exactly one less edge than H. Define weights on the cover relations as follows.

Let u(G,H) be the number of subsets non-edges of G which, when added to G, result in H. This corresponds to upward construction.

Let d(G,H) be the number of edges of H which, when deleted, result in G. This corresponds to the downward deconstruction [26].

The poset is shown in Fig. 3.25.

It is a complexification of the initial 4 dots frame.

In this poset, the weight of a Hasse walk of shape W from G to H will represent the number of ways to transform G to H through a sequence of edge additions and deletions specified by W.

Figure 3.26 shows an example of commutation condition for graphs spanning [26].

The negative result is related to the area we selected to apply DU-UD operation.

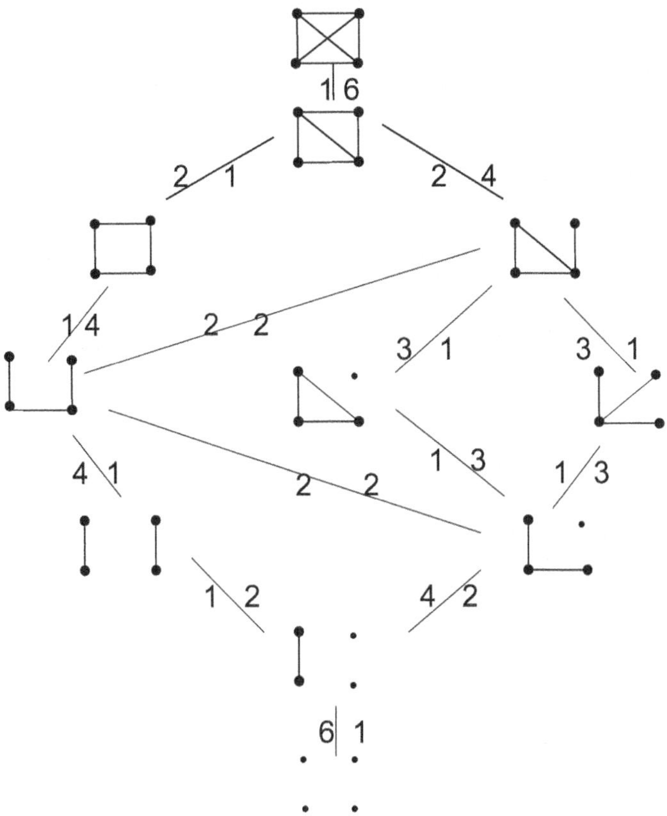

Fig. 3.25 Poset for complexification

Fig. 3.26 Commutation condition for graphs spanning

There are areas of increasing and areas of decreasing complexification.

The poset for complexification is a 2D table.

As for dual graded graphs, DGG we can develop Polytopic Roadmaps associated to the graph spanning.

We denote by U the "up" construction and by D the "down" deconstruction as illustrated in Fig. 3.25.

Despite the fact that we refer to the same poset, the U and D evolutions represent the doubling or twinning process.

The method to increase dimensionality starting from dual graphs is illustrated in Fig. 3.27. The starting diagram is a U-graph (2D) or a pair U, D-graphs as shown in

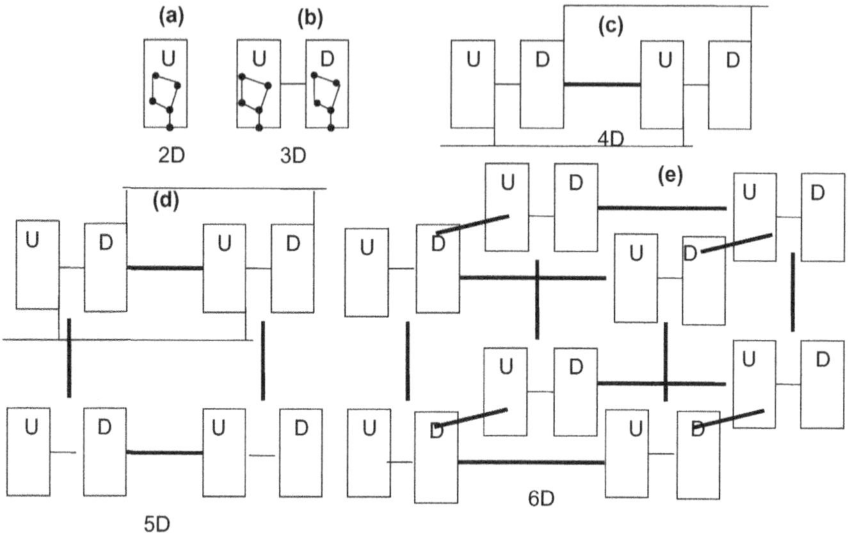

Fig. 3.27 Polytopic Roadmaps for graphs spanning

Fig. 3.27a. The pair U, D results from U poset by doubling followed by relating. It is a 3D polytope, as shown in Fig. 3.27b. The relating is indicated here by a thin line U-D. Relation between U and D posets may be that from commutation operation, namely simple translations or may be different from this.

The next doubling increase dimension to 4D. A new relation is shown by a thick line in Fig. 3.27c. The previous relation between U and D is maintained but new relation should be considered between the two cubes U-D. A proposal in which only U graphs or D graphs are involved is shown in Fig. 3.27c.

Continuing the doubling and relating, we obtain the 5D frame from Fig. 3.27d and then the 6D frame from Fig. 3.27e. The relating is different for different dimensions.

3.4.2 Polytopic Chemistry

The dimension increase for chemicals according to the sequence: molecular-polymers-supramolecular-reticular-polytopic chemistry is presented here as another example of complexification.

Chemists have focused on molecules considered as (0D) and to some extent on polymers considered as (1D) but beyond these the ability to express control in 2D and 3D remained less developed.

Figure 3.28 illustrates possibly Polytopic Roadmaps from molecular to polytopic chemistry. The label polytopic use to be associated to four and more dimensions taking into account that 0D, 1D, 2D and 3D has conventional names [3, 41].

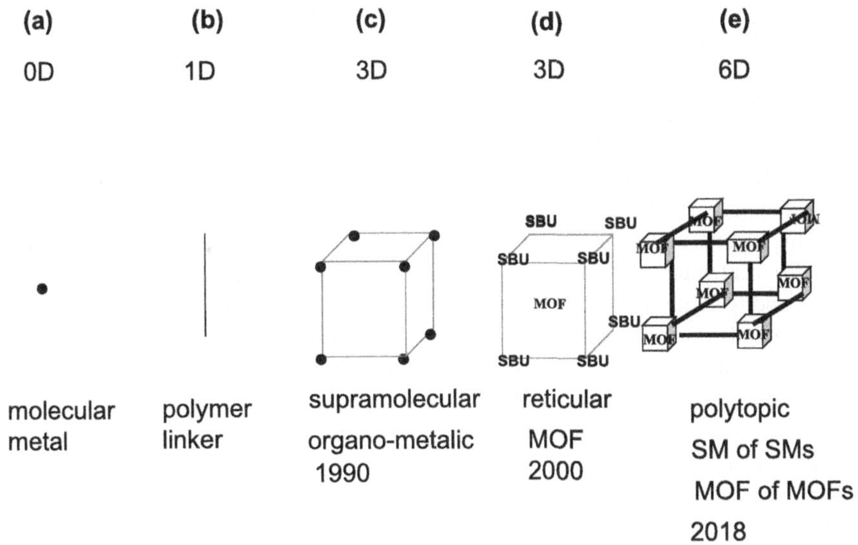

Fig. 3.28 Polytopic Roadmaps for chemistry

Figure 3.28a shows an atom, 0D, in this case a metal, and Fig. 3.28b a polymer 1D that may be used as linker for the metal atoms.

The question of how molecules interact was addressed by molecular chemistry and by supramolecular chemistry [19]. Supramolecular chemistry is based on weak bonds between atoms and molecules. The study of self-assembling systems in chemistry has delivered a wide variety of discrete molecular architectures, as for instance twisted structures (helicates), latticed structures (grids, racks, ladders), filamentous structures (rods, metallodendrimers), and interlaced structures (rotaxanes, catenanes, knots). Interesting architectures are to be found in the closed, geometrically-shaped, 2D and 3D structures known as metallocyclic polygons and polyhedra. These materials have been variously termed molecular triangles, squares, hexagons, cubes, and so on, because of their resemblance to the corresponding geometric forms. Such molecules typically contain central cavities, allowing them to act as artificial molecular scale containers or receptors. In doing so, they may exhibit unusual electrochemical, magnetochemical, photoluminescent, catalytic, or synthetic chemical effects.

In the early of 1990 pioneering work has been done concerning the supramolecular organo-metalic compounds [10]. The "node and spacer" approach, to guide the design of porous coordination polymers was applied. Compounds as that shown in Fig. 3.28c have been studied [30, 31]. In this case the approach was "metal and ligand". The result was referred as organo-metalic supramolecules (SM).

With the advent of reticular chemistry the interactions between molecules were completed to make strongly bonded frameworks, which can withstand multiple organic reactions [20, 38]. Reticular chemistry compound shown in Fig. 3.28d replaced the metal from SM organo-metalic shown in Fig. 3.28c by SBUs (secondary building units) [16].

The metal-organic frameworks, (MOF), consist of both inorganic and organic units. The organic units (linkers/bridging ligands) consist of carboxylates, or anions, such as phosphonate, sulfonate, heterocyclic compounds and so on. The inorganic units are the metal ions or the clusters termed as SBUs.

MOF geometry is determined by the coordination number, coordination geometry of the metal ions, and the nature of the functional groups. A variety of SBU geometries with different number of points of extension such as octahedron, trigonal prism, square and triangle have been studied as MOF structures. A bridging ligand reacts with a metal ion with more than one vacant or labile site. The final framework topology of MOF is determined by both SBU connectors and organic linkers.

Reticular chemistry started with linking inorganic clusters into extended porous frameworks, and then linking organic molecules and metal ions into metal-organic frameworks (MOFs), and organic molecules together into covalent organic frameworks (COFs). It opened new ways for developing chemistry beyond the molecule.

As a result of reticular chemistry development, chemists begin to consider the framework as a molecule.

When chemists refer to all dimensions they use to limit the exam to 0D, 1D, 2D and 3D [37]. These dimensions are conventional. They characterize the pattern or skeleton of the framework.

But the chemistry is not limited to 3D. We need to consider new stages in chemistry development. After molecular, polymeric, supramolecular, reticular chemistry we need to consider polytopic chemistry.

The polytopic architectures look to four and higher than four dimensions [12, 13].

The polytopic chemistry may result by iteration method expressing the development from molecule to molecule of molecule, (as it is SM), from SM to SM of SMs or from MOF to MOF of MOFs and so on. Figure 3.28e shows an example of MOF of MOFs. It results by replacing the SBU from MOF by other MOFs. The iteration process endorses higher dimensionality. Obviously the new compounds assimilation is accompanied by an accommodation process, relating molecules in the new framework. This is a chemical counterpart of the Piagetian process of assimilation and accommodation.

The polytope shown in Fig. 3.28e is a representation known as 6D cube and consisting from six 3D cubes. The 6D cube is obtained by joining all corners of a given 3D cube with the corresponding corners of the neighboring 3D cubes [1, 33].

The eight lines connecting two 3D cubes associated to MOFs, are represented in Fig. 3.28e by thick hyper-lines. In this way the 6D cube has a simplified representation. Using the thick hyper-line the 6D cube is shown in Fig. 3.28e as a cube of cubes or in other terms a MOF of MOFs.

Figure 3.29 shows the Periodic Table for MOFs-variant 1.

The columns correspond in succession to the molecule (dots), polymer (lines), MOF, (cubes). It is assumed that the lines self-associate in cubes or in chemical terms polymers self-associate in MOFs.

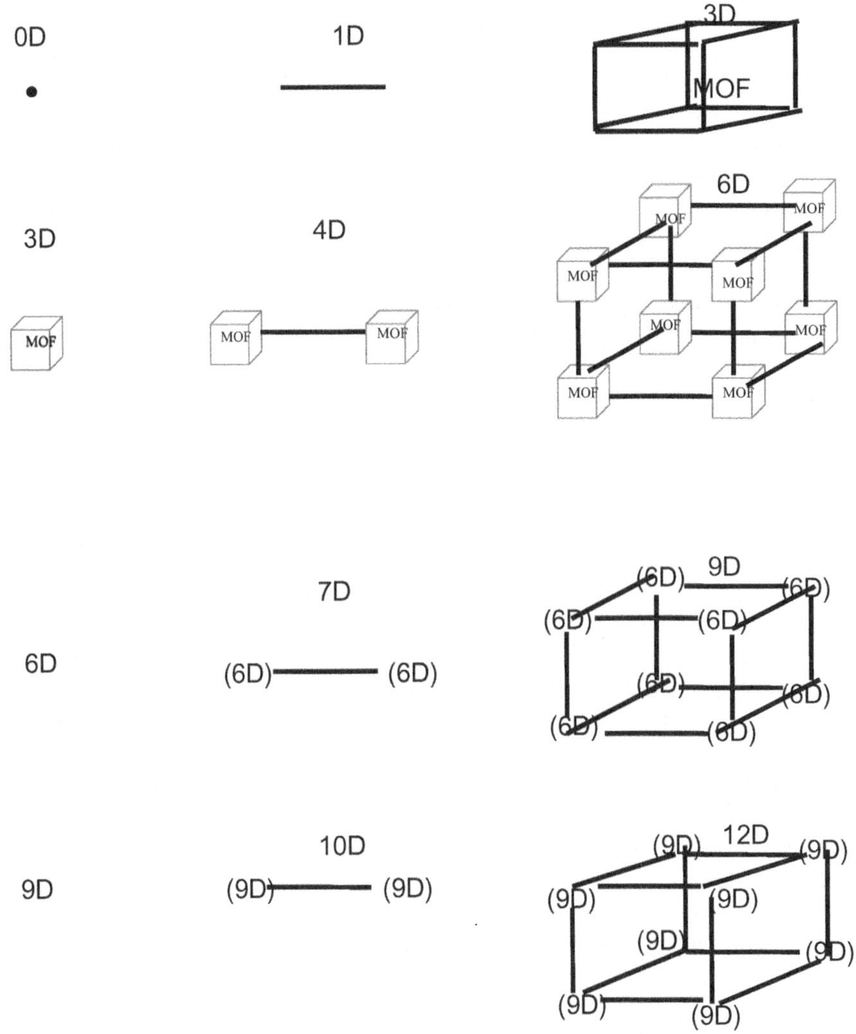

Fig. 3.29 Periodic Table for MOFs-variant 1

The first row starts from molecules and finish with MOF represented as 3D-cube.

The second row starts from MOF cube and finish with MOF of MOFs represented as 6D-cubes or cube of cubes.

The third row starts from 6D-cubes and finish with 9D-cubes (MOF of MOF of MOFs).

The fourth row, starts from 9D-cubes and finish with 12D-cubes.

The fifth row, omitted in this figure, will starts from 12D-cubes and will finish with 15D-cubes.

Rows correspond to the periods of the Periodic Table.

The thick-lines represent the multiple linkers.

Columns show the groups of the table. It is expected that the MOFs in the same column will show some common properties related to the basic pattern or skeleton: dot, line or cube.

The next row unincluded in this figure will show 12D, 13D and 15D frames.

All the frameworks included in the Periodic Tables should be considered as molecules.

The molecule fixes the atom in a specific geometry and spatial arrangement, and likewise, the polytopic framework fixe the molecules in a specific geometry and spatial arrangement. The uniqueness of the polytopic framework is that its chemistry is significantly richer than its constituent molecules because it encompasses high dimensional spaces into which matter can be further manipulated and controlled. More dimensions support more capabilities. Polytopic Roadmaps represents the natural progression of the molecules and reaction mechanisms.

As a new paradigm, polytopic chemistry should be submitted to the applications test.

Certain self-evolving chemical systems and some phase transitions can best be understood in higher dimensions. Verifying such hypotheses could improve our understanding of phase transitions–area critical to the pharmaceutical and bio-medical industry, catalysis and material sciences. If we can understand what happens during a phase transition in higher dimensions we will gain a better understanding for self-evolvable smart or intelligent materials. Gaining this information raises the question of whether it is possible to build molecular structures that are controlled by design. If we can design the structure, it follows that we will also be able to tailor the physical and chemical properties to match specific properties or needs.

While progress has been made toward rational design of molecules, SM or MOF significant challenges remain, particularly with respect to multiple lengths or temporal scales or to self-evolvable smart systems. Even so it is possible that well-tailored polytopic chemicals with desired properties will be rationally designed in the near future and the exploration of their practical applications will be a stimulating scientific activity.

References

1. Ammon, J.: Hypercube Connectivity within ccNUMA Architectures, SGI Origin Team, Mountain View, CA (1998)
2. Cerf, L.: Constraint-Based Mining of Closed Patterns in Noisy n-ary Relations. Ph.D. Thesis, INSA-Lyon, July (2010)
3. Coxeter, H.S.M.: Regular Polytopes, 3rd edn. Dover, New York (1973)
4. Dau, F., Wille, R.: On the modal understanding of triadic contexts. In: Decker, R., Gaul, W. (eds.) Classification and Information Processing at the Turn of the Milenium, Proc. Gesellschaft für Klassifikation (2001)
5. Fomin, S.: Dual graphs and Schensted correspondences. In: Leroux, P., Reutenauer,C., (eds.), Series Formelles Et Combinatoire Algebrique., pp. 221–236. Montreal, LACIM, UQAM (1992)
6. Fomin, S.: Duality of graded graphs. J. Algebraic Combin. **3**, 357–404 (1994)

7. Fulton, W.: Young Tableaux: With Applications to Representation Theory and Geometry. Cambridge University Press, Cambridge (1997)
8. Ganter, B., Wille, R.: Formal Concept Analysis. Mathematical Foundations. Springer, Berlin (1999)
9. Hoffman, M.E.: Updown categories and algebraic structures. AMS Special Session on Algebraic and Enumerative Combinatorics, Joint Mathematics Meetings, San Antonio, Texas (2006)
10. Hoskins, B.F., Robson, R.: Design and construction of a new class of scaffolding-like materials comprising infinite polymeric frameworks of 3D-linked molecular rods. J. Am. Chem. Soc. **112**(4), 1546–1554 (1990)
11. Iordache, O.: Polystochastic Models for Complexity. Springer, Berlin, Heidelberg (2010)
12. Iordache O.: Self-evolvable systems. Machine learning in social media. Springer, Berlin, Heidelberg (2012)
13. Iordache, O.: Advanced Polytopic Projects. Springer, Cham, Switzerland (2018)
14. Iordache, O., Corriou, J.P., Garrido-Sanchez, L., Fonteix, C., Tondeur, D.: Neural network frames applied for biochemical kinetic diagnosis. Comp. Chem. Engng. **17**, 1101–1113 (1993)
15. Iordache, O., Corriou, J.P., Tondeur, D.: Separation sequencing. use of information distance. Canad. Journ. Chem. Engng. **71**(12), 955–966 (1993b)
16. Kalmutzki, M.J., Hanikel, N., Yaghi, O.M.: Secondary building units as the turning point in the development of the reticular chemistry of MOFs. Sci. Adv. **4**(10), eaat9180 (2018)
17. Kingman. J.F.C.: Random partitions in population genetics. Proceedings of the Royal Society of London. Series A, Mathematical and Physical Sciences, **361**(1704), 1–20 (1978)
18. Lehmann, F., Wille, R.: A triadic approach to formal concepts analysis. In: Ellis, G., Levinson, R., Rich, W., Sowa, J.F.: (eds.) Conceptual Structures: Applications Implementations and Theory, vol. 954, Lectures Notes in Computer Science, Springer, (1995)
19. Lehn, J.-M.: Supramolecular chemistry: Concepts and perspectives. Wiley-VCH, Weinheim (1995)
20. Li, H., Eddaoudi, M., O'Keeffe, M., Yaghi, O.M.: Design and synthesis of an exceptionally stable and highly porous metal-organic framework. Nature **402**(6759), 276–279 (1999)
21. Linial, N., Luria, Z.: An upper bound on the number of high-dimensional permutations. Combinatorica **34**(4), 471–486 (2014)
22. Lounkine, E., Bajorath, J.: Adaptation of formal concept analysis for the systematic exploration of structure-activity and structure-selectivity relationships. J. Cheminformatics, **2**(suppl 1):21 (2010)
23. Muralidharan, V.T., Rajan, B.S.: Physical layer network coding for the k-user multiple access relay channel. IEEE Trans. Wireless Commun. **12**(6), 3107–3119 (2013)
24. Mustonen, V., Rajesh, R.: Numerical estimation of the asymptotic behaviour of solid partitions of an integer. J. Phys. A: Math. Gen. **36**(24), 6651 (2003)
25. Negri, E., Fumagalli, L., Macchi, M.: A review of the roles of digital twin in cps-based production systems. Proc. Manuf. **11**, 939–948 (2017)
26. Sloss, C.A.: Enumeration of walks on generalized differential posets. M.S. Thesis, Univ. of Waterloo, Canada (2005)
27. Stanley, R.: Differential posets. J. Amer. Math. Soc. **1**, 919–961 (1988)
28. Stanley, R.P.: Variations on differential posets. Invariant theory and tableaux. IMA Vol. Math. Appl. **19**, 145–165. Springer, New York (1990)
29. Stanley, R.P.: Enumerative Combinatorics, vol. 2. Cambridge University Press, Cambridge (1999)
30. Swiegers, G.F., Malefetse, T.J.: New self-assembled structural motifs in coordination chemistry. Chem. Rev. **100**(9), 3483–3538 (2000)
31. Swiegers, G.F., Malefetse, T.J.: Classification of coordination polygons and polyhedra according to their mode of self-assembly. 2., Review of the literature. Coordination chemisstry reviews, **225**(1–2), 91–121 (2002)
32. Tao, F., Qi, Q., Wang, L., Nee, A.Y.C.: Digital twins and cyber–physical systems toward smart manufacturing and industry 4.0: correlation and comparison. Engineering, **5**(4), 653–661 (2019)

33. Thiel, T.: The Design of the Connection Machine, Design Issues, **10**, 1, 5–18, Spring (1994)
34. Tibbits, S.: 4D printing XE "printing": multi-material shape change. Architect. Des. **84**(1), 116–121 (2014)
35. Valtchev, P., Missaoui, R.: Similarity-based Clustering versus Galois lattice building: Strengths and Weaknesses. ECOOP'2000 Workshop on Objects and Classification: a Natural Convergence, June 2000, Sophia-Antipolis, France (2000)
36. Voutsadakis, G.: Polyadic concept analysis. Order **19**, 295–304 (2002)
37. Yaghi, O.M.: Reticular chemistry in all dimensions. ACS Central Sci. **5**(8), 1295–1300 (2019)
38. Yaghi, O. M., Li, H., Davis, C., Richardson, D., Groy, T. L.: Synthetic strategies, structure. patterns, and emerging properties in the chemistry of modular porous solids. Acc. Chem. Res., **31**(8), 474–484 (1998)
39. Yong, A.: What is a young tableau. Notices of the AMS **54**(2), 240–241 (2007)
40. Wille, R.: The basic theorem of triadic concept analysis. Order **12**, 149–158 (1995)
41. Ziegler, G. M.: Lectures on polytopes. graduate texts in mathematics, 152. Springer-Verlag, New York (1995)
42. Wollbold, J., Huber, R., Kinne, R., Wolff, K.E.: Conceptual Representation of Gene Expression Processes. In: Wolff, K.E. et al. (eds.) Knowledge Processing and Data Analysis. LNAI **6581**, pp. 77–99. Springer-Verlag, Heidelberg (2011)
43. Wolff, K.E.: Temporal Concept Analysis. In: Mephu Nguifo, E. et al. (eds.): ICCS-2001 International Workshop on Concept Lattices-Based Theory, Methods and Tools for Knowledge Discovery in Databases, Stanford University, Palo Alto (CA), pp. 91–107 (2001)
44. Wolff, K.E.: States, transitions, and life tracks in temporal concept analysis. In: Ganter, B., Stumme, S., Wille, R.: (eds.) Formal Concept Analysis—Foundations and Applications, LNAI **3626**, pp 127–148 Springer, Heidelberg (2005)
45. Wolff, K.E.: Applications of temporal conceptual semantic systems. In: Wolff, K.E., et al. (eds.) Knowledge Processing and Data Analysis. LNAI **6581**, pp. 60–76. Springer-Verlag, Heidelberg (2011)

Chapter 4
Equipments

4.1 Multi-scales

Chemical processes engineering are undergoing rapid changes facing the challenges of tightening energy and resources, climate change and high complexity advent. Process intensification that leads to smaller, less costly, cleaner, safer, higher productivity and energy efficient technology was proposed as a new paradigm for chemical process engineering[21, 22].

The innovative design of high yield processes or compact equipments has become one of the pressing industrial needs in recent years.

One of the methods for process intensification is the use of equipments with locally miniaturized structures, that is, micro-reactors or mini-channels because of their enhanced momentum, heat and mass transfer properties. Miniaturized process and energy equipments can be heat exchangers, chemical mixers, distributors and collectors, micro-reactors or fuel cells [6, 7].

To obtain a comparable productivity with that of conventional equipment, a number of micro-reactors or mini-channels should be installed in parallel and series.

The numbering-up and numbering-down processes are key issues for large-scale industrial applications of miniaturized equipments [3, 4, 13]. Numbering-up and numbering-down are distinctive scale-up or scale-down methods.

The fluid distribution uniformity among the parallel channels has an important role on the global performance of multi-channel equipments.

This is obvious when multi-scale ladder-type fluid networks are involved [2, 15, 16]. In order to achieve uniform flow distribution among all parallel channels in the network, a basic step is the design of an optimized two-scale elementary Z-type ladder circuit. In such elementary fluidic circuit, the single inlet port and single outlet port are located on opposite sides of the bundle of parallel cross-channels, meaning that the flow direction is the same in the distributor and the collector pipes. All cross-channels are assumed to have the same geometrical characteristics so that the passage-to-passage irregular distribution may be neglected. On the contrary, the

O. Iordache, *Polytopic Roadmaps*, Studies in Systems, Decision and Control 368, https://doi.org/10.1007/978-3-030-75630-7_4

improper design of fluid distributor and collector pipes configuration is the main cause of flow irregular distribution among the parallel cross-channels.

Several studies have then been focused on how to improve the flow distribution uniformity of the elementary Z-type ladder circuit. A relatively uniform distribution may be approached by making their hydraulic resistance much larger than that of the distributor and collector pipes if the latter have a uniform profile [18, 19]. This implies that the distributor and collector pipes are large and hindering, which is unfavorable for miniaturized equipments. Instead of the uniform profile, rectangular shape of the distributor and collector, alternative shapes have been proposed.

Multi-scale transfer equipments useful for process intensification, distributors and collectors and equipments for cyclic operations have been characterized using Polytopic Roadmaps and classified in Periodic Tables.

One of the main objectives of the study is to outline and implement the isomorphism between the cognitive-intelligence development stages, and the smart equipments development stages.

Figure 4.1 illustrates the development of Polytopic Roadmaps designed for multi-scale transfer. Observe that the dimension increase is based on doubling or twining with relating.

The main module may be assimilated to a cube as shown in Fig. 4.1d. It results from previous stages illustrated in Fig. 4.1 dots (0D), lines (1D), squares or rectangles (2D) by doubling and relating.

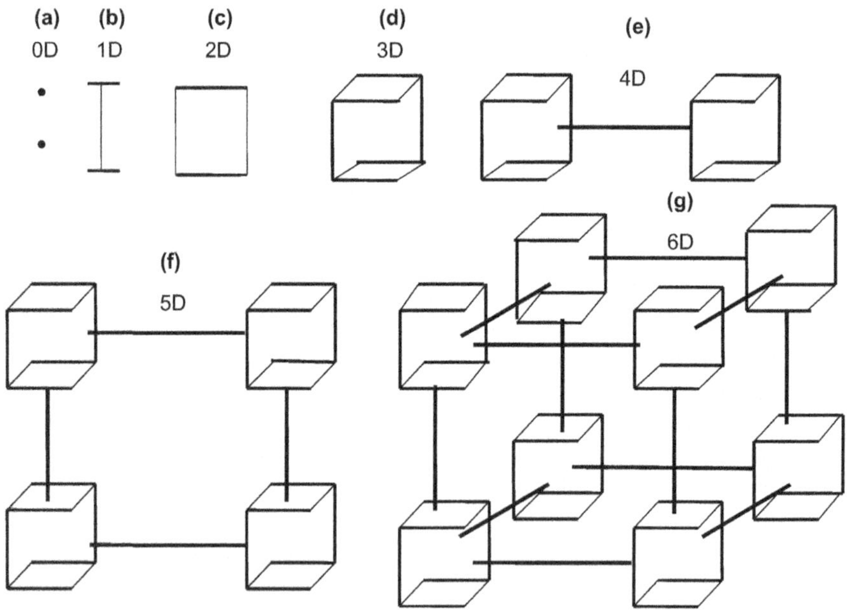

Fig. 4.1 Polytopic Roadmaps for multi-scale transfer

The polytope shown in Fig. 4.1e is a representation known as hypercube or 4D cube and consisting from a pair of 3D cubes. The 4D cube is obtained by joining all corners of the first 3D cube with the corresponding corners of the second 3D cube. The eight lines connecting the two 3D cubes are represented in Fig. 4.1e by a thick hyper-line. In this way the 4D cube has a simplified representation. All dimensions higher than three may be represented using thick hyper-lines in order to visually simplify the resulting structures [1, 24]. Using the thick hyper-line the 4D cube is shown in Fig. 4.1e as a line of cubes. The dimension is $1 + 3 = 4$. In this sum 1 corresponds to the line and 3 to the cubes.

Figure 4.1f shows a 5D cube. Figure 4.1g shows a 6D cube. Using the thick hyper-lines, the 5D cube is shown as square of cubes of dimension $2 + 3 = 5$, while the 6D cube is shown as cube of cubes of dimension $3 + 3 = 6$.

In practice the interconnection between cubes may be incomplete.

Figure 4.2 Shows polytopic Roadmaps for multi-scale transfer.

The incomplete polytope 4D for multi-scale transfer is the result.

In this case just two connections between the two cubes of the 4D equipment have been considered. Despite incompleteness the devices may ensure an efficient transfer. Obviously more connections imply higher versatility and smartness potentialities for transfer equipments.

Figure 4.2 illustrates the increasing of dimension from 0D to 4D.

To clarify the proposed 4D design, Fig. 4.3 shows more details for incomplete

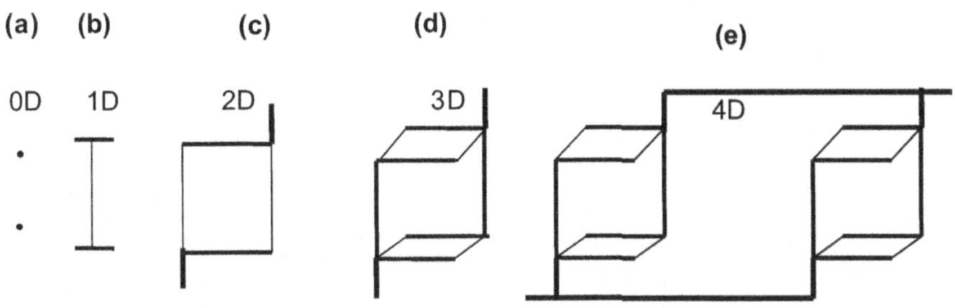

Fig. 4.2 Polytopic Roadmaps: 4D for multi-scale transfer

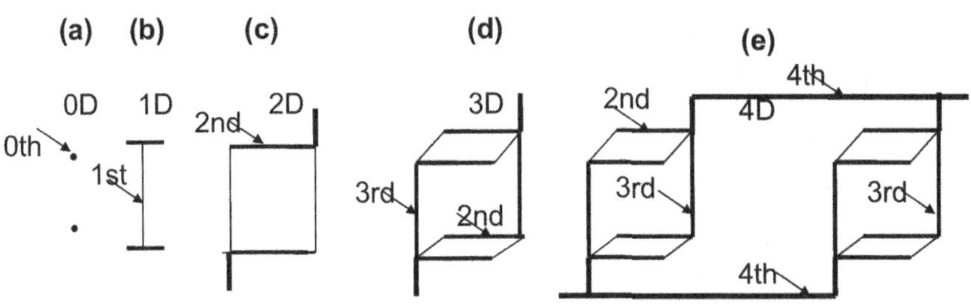

Fig. 4.3 Polytopic Roadmaps: 4D for multi-scale transfer-details

polytope 4D multi-scale transfer. The scales from 0 to 4th are considered here. They are correlated to fluid flows.

Figure 4.3a shows dots and illustrates the absence of transfer that is 0th scale or no scale available for transfer.

Figure 4.3b shows a line, that is one micro-channel and the 1st scale of transfer.

Figure 4.3c shows several micro-channels. They are in the same plane. Only two have been shown here but a plane network, a square or rectangle may be considered.

Figure 4.3d shows several parallel micro-plates. They are the top and bottom faces of the cube. This cube is the building block of the transfer equipment. Several plates may be included but the equipment is 3D. The 3rd scale of transfer is included here.

Figure 4.3e shows one micro-equipment. It results by interconnecting the 3rd scale network.

The thick hyper-line connecting the two cubes in 4D device represents the eight connections between corresponding vertices of the cubes. This is the ideal 4D situation as shown in Fig. 4.3e. Less than eight connections will be considered in practice.

These correspond to incomplete 4D equipments.

Figure 4.4 show distinct incomplete polytopes 5D and 6D for multi-scale transfer.

Figure 4.5 shows more details for incomplete polytope 5D and 6D dedicated to multi-scale transfer.

Figure 4.5a, b shows micro-equipments resulting from 5 and 6D geometry.

They result by simplifying the interconnections specific to complete 5D and 6D geometry. These geometries have been extensively studied as parallel computer frameworks [1, 20].

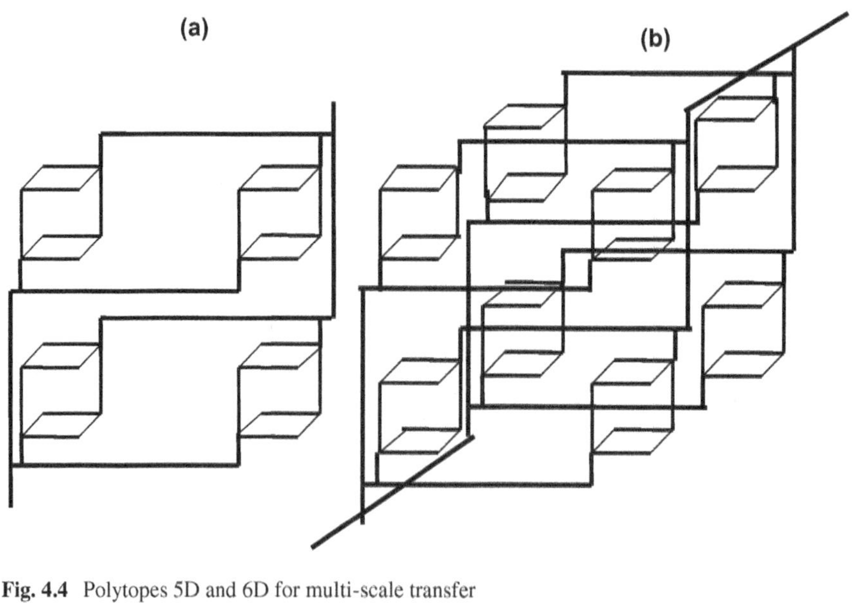

Fig. 4.4 Polytopes 5D and 6D for multi-scale transfer

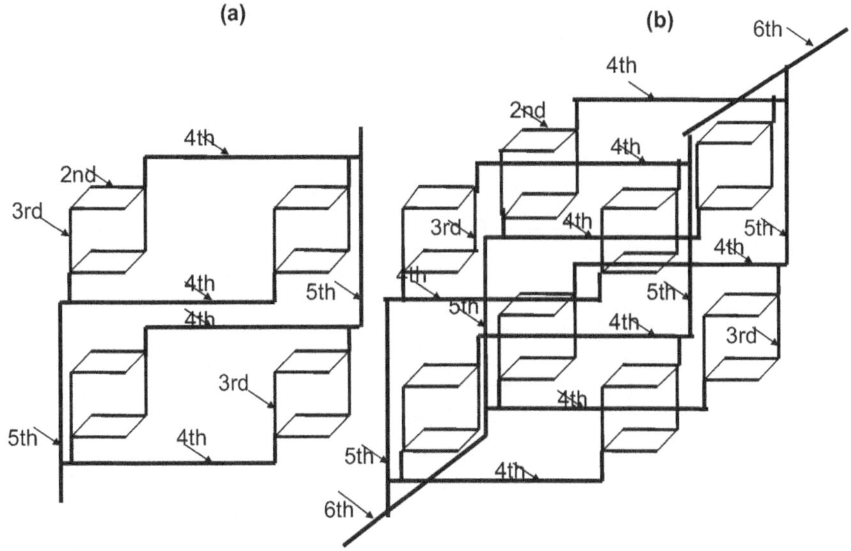

Fig. 4.5 Polytopes 5D and 6D for multi-scale transfer-details

For heat transfer domain, Fig. 4.5 illustrates how heat pipes or heat conducting rods may be combined to move the heat into or out of a poorly conducting structure.

Figure 4.6 shows a Periodic Table for multi-scale equipments.

The columns correspond in succession to: lines, rectangles, and cubes.

The first row of the table starts from one line and finishes with a 3D-cube.

The second row starts from the line relating two 3D-cubes (4D-cube) and finish with cube of cubes represented as 6D-cubes of dimension $3 + 3 = 6$.

The third row starts from line the relating two 6D-cubes (7D-cube) and finish with 9D-cubes (cube of cubes of cubes) of dimension $3 + 3 + 3 = 9$.

The next row unrepresented in Fig. 4.6 will correspond to 10D, 11D and 12D equipments.

The thick-lines represent the interconnections between the cubes.

Any equipment from the Periodic Table may be integrated from design stage in the installation that controls the transfer. It plays the role of scaffold or skeleton for specific transfer equipment. The versatility in functioning is obtained by activation or inactivation of the interconnections. The model of fluid flow through the transfer equipment appears as a digital twin DT of the real fluid flow model in the installation that needs the transfer. Both the real model and the virtual model support the equipment design as a complex product [12]. The DT creates high-fidelity virtual models of physical objects in virtual space in order to simulate their behaviors in the real world and provide feedback [23].

Figure 4.7 shows the Polytopic Roadmaps for distributors and collectors systems.

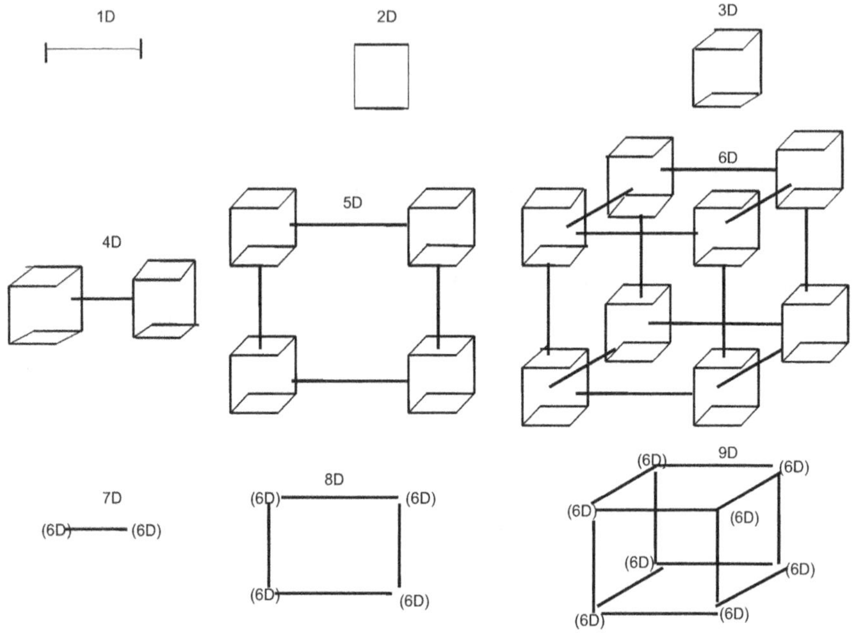

Fig. 4.6 Periodic Table for multi-scales equipment

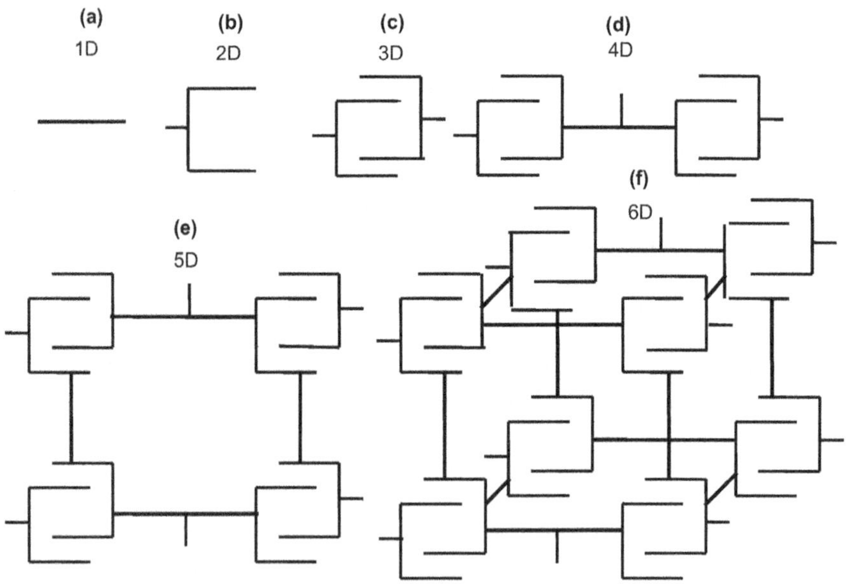

Fig. 4.7 Polytopic Roadmaps for distributors and collectors

Figure 4.7 illustrates the development of Polytopic Roadmaps designed for multi-scale distribution and collection. Observe that the dimension increase is based on doubling or twining with relating.

Figure 4.7b is a 2D distributor. Fluid enters the distributor through a single inlet flow through the branching channels and leaves the distributor through a square array of outlets. This distributor could uniformly lead air over the catalytic layer of a fuel cell as well as distribute water. Such structures could feed high-throughput setups or uniformly heat or cool surfaces. They may be integrated into micro-fluidic devices [5].

The main module may be assimilated to a cube as shown in Fig. 4.7c. It includes the distributor and the collector in a 3D frame. The connection between distributor and collector is completed by the fluid surrounding these devices.

The polytope shown in Fig. 4.7d is a hypercube or 4D cube consisting from a pair of 3D cubes. The 4D cube is obtained by joining corners of the first 3D cube with the corresponding corners of the second 3D cube. The lines connecting the two 3D cubes are represented in Fig. 4.7d by a thick hyper-line. In this way the 4D cube has a simplified representation.

Using the thick hyper-line the 4D cube is shown in Fig. 4.7d as a line of cubes. The dimension is $1 + 3 = 4$. In this sum 1 corresponds to the line and 3 to the cubes.

Figure 4.7e shows a 5D cube. Figure 4.7f shows a 6D cube. Using the thick hyper-lines, the 5D cube is shown as square of cubes of dimension $2 + 3 = 5$, while the 6D cube is shown as cube of cubes of dimension $3 + 3 = 6$.

In practical cases the interconnection between building cubes may be incomplete.

Figure 4.8 shows a Periodic Table for multi-scale distributors and collectors.

The columns correspond in succession to: lines, squares or rectangles, and cubes.

The first row of the table starts from one line and finishes with a 3D-cube.

The second row of the table starts from the line relating two 3D-cubes (4D-cube) and finishes with cube of cubes represented as 6D-cubes of dimension $3 + 3 = 6$.

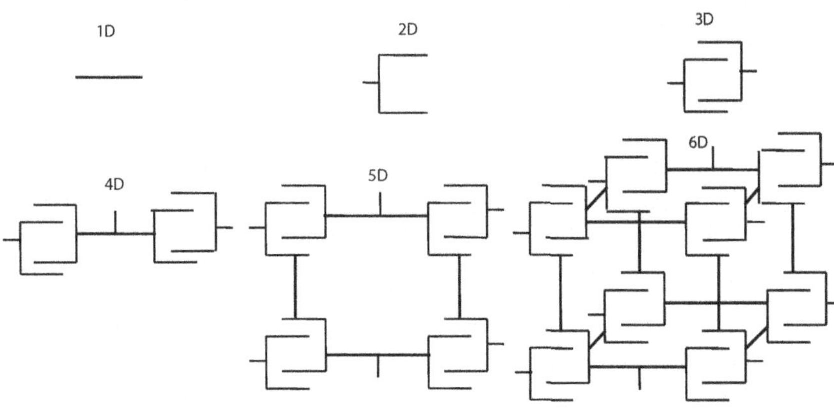

Fig. 4.8 Periodic Table for distributors and collectors

4.2 Cyclic Operations

In chemical industries, operating modes as reversed flow for reaction-regeneration energy efficient coupling of endothermic and exothermic reactions, countercurrent flow and induced pulsing flow in trickle beds, unsteady operations, cyclic processes, extreme conditions, low-frequency vibrations to improve gas–liquid contacting in bubble-columns, high temperature and high pressure technologies, and supercritical media, and use of composite structured packing achieving low pressure drop through vertical stacking of catalyst, have been considered for practical application [17].

Cycling operation methods are of great importance in oil chemistry, in pharmaceutical and food industry, bio-refinery, isotopes separation, hydrogen purification, desalinization, and so on [26]. Cyclic separation technologies such as pressure swing adsorption (PSA), temperature swing adsorption (TSA), vacuum swing adsorption, cyclic zone adsorption, simulated moving beds (SMB) chromatography, pressure swing reactor and reverse flow reactor, parameter pumping and so on, are unsteady non-linear processes difficult to put into practice and to control [15].

In pressure swing adsorption (PSA) processes, gas mixtures are separated by selective adsorption over a bed of sorbent materials. The cyclic nature of these processes arises from the high pressure adsorption phase and the subsequent low-pressure regeneration phase. The PSA cycle was accepted for commercial use in air drying.

Thermal swing adsorption (TSA) processes are similar to pressure swing adsorption processes and also separate gas mixture, but here the cyclic nature of these processes arises from the low temperature adsorption phase and the subsequent high temperature regeneration phase. There exist processes that are a combination of PSA and TSA.

A reverse flow reactor is a packed bed reactor in which the flow direction is periodically reversed in order to trap a hot zone within the reactor. In this way even systems with a small adiabatic temperature rise can be operated without preheating the feed stream.

In a pressure swing reactor, reaction and adsorption occur in the same bed.

The adsorption is typically used to purify one of the reaction products. The cyclic nature of a pressure swing reactor arises from the same high pressure adsorption and low pressure regeneration phases as in the pressure swing adsorption.

As the separation complexity and the number of devices increases it becomes very difficult to formulate a feasible schedule of the cyclic operation.

The proposed approach allowing operating cyclic separations in multi-component high complexity conditions is that of self-evolvable and smart separation systems. These are systems that can change autonomously the schemes and the dynamic behavior and are capable to control and to take advantage of the unexpected events of their environment in increasingly complex ways. Smart separation systems should have emergent, not entirely pre-programmed, behavior.

Non-stationary and periodically operated separation device with self-evolvability based on schemes modification on self-configuring schemes and multi-scale schemes organized by self-similar replication at different conditioning levels are significant.

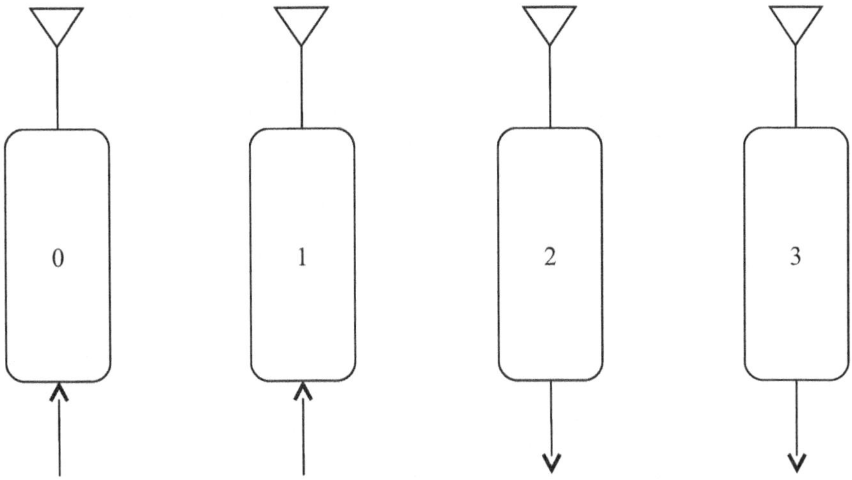

Fig. 4.9 Four steps scheme

Figure 4.9 shows a four steps scheme.

Well-known examples are the scheme involving 4-beds and 4-steps [15, 17, 26].

A sequence of the above described steps is called a cycle. That cycle is a 2D presentation of the scheme.

The operation steps are denoted as follows:

0-compression

1-production

2-decompression

3-purge.

For the reverse or modified process the notations are:

0'-modified compression

1'-modified production

2'-modified decompression

3'-modified purge.

Figure 4.10 shows a two-scale PSA scheme.

Using the thick hyper-lines, the 5D cube is shown as square of cubes.

The main steps denoted **0**, **1**, **2**, and **3** may be subdivided by a geometric lower scale having four sub-steps. This may be considered as separation schemes of separation schemes.

For instance, 01 means the step **0** at larger scale but 1 at the smaller scale. The step 01' reverses or modifies the step 01.

Figure 4.11 shows a 6D polytope for two-scale PSA scheme.

With the thick hyper-lines, the 6D cube is shown as cube of cubes.

A conservative development strategy was selected for Fig. 4.11.

The main steps **0**, **1**, **2**, and **3** are modified to the main steps **0'**, **1'**, **2'**, and **3'**.

Fig. 4.10 Polytope 5D for PSA separation

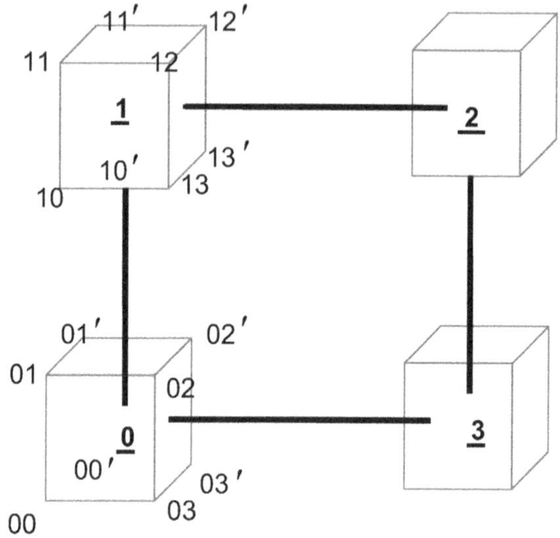

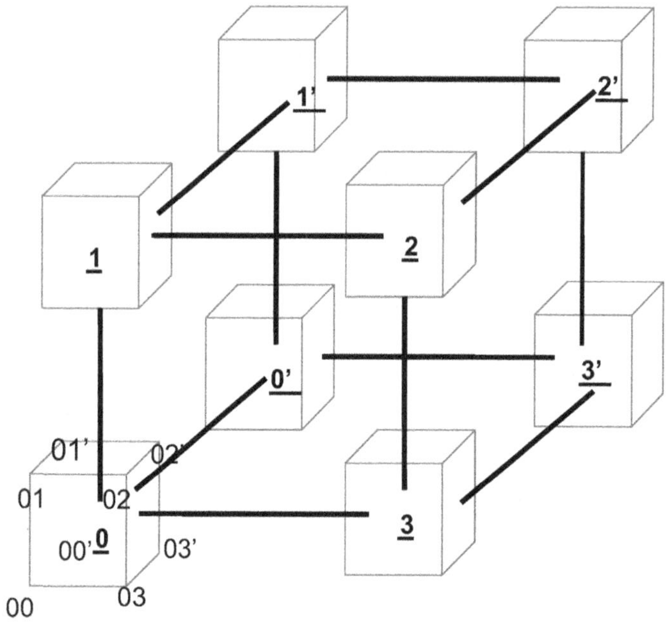

Fig. 4.11 Polytope 6D for PSA separation

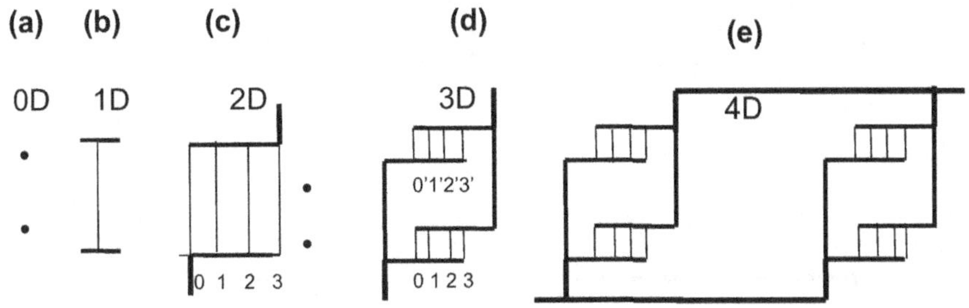

Fig. 4.12 Polytopic Roadmaps: 4D for cyclic operations-details

An exploratory development where after the main steps **0**, **1**, **2**, and **3** new main steps **4**, **5**, **6**, and **7** arise may be of interest. The 4-beds and 4-steps process would be replaced by an 8-beds and 8-steps process.

For instance we may couple the 4-beds pressure swing absorption PSA with 4-beds temperature swing absorption TSA. Direct sequence operates PSA while reverse sequence operates TSA.

Figure 4.12 shows details for the Polytopic Roadmaps for 4D for cyclic operations.

It offers details about 4D but helps to understand what means high-dimensional cyclic separation.

In this case the four steps process has been considered. The notations are 0, 1, 2 and 3:

For modified process the notations are 0', 1', 2' and 3'.

Figure 4.12a illustrates the absence of adsorption that is 0th scale or no scale available for processing.

Figure 4.12b shows just one column and the 1st dimension of process.

Figure 4.12c shows four columns in the stage 0, 1, 2 and 3. They are represented in the same plane. The 2nd dimension of process is presented.

Figure 4.12d shows 2 parallel plates. They are the top and bottom face of the cube. Several plates may be included but the device is 3D. The 3rd dimension of process is presented here.

Figure 4.12e shows one device resulting by interconnecting of the 3rd dimension network of two cubic devices. The 4th dimension of the process is presented here.

4.3 Conveyors

As the traditional manufacturing combines with information technology, organizations are tapping into a level of technical orchestration never attainable before. Systems of systems, SoS, facilitate real-time interactions of people, machines, assets, systems, and things. An illustration is the Smart Factory; the factory ecosystem of

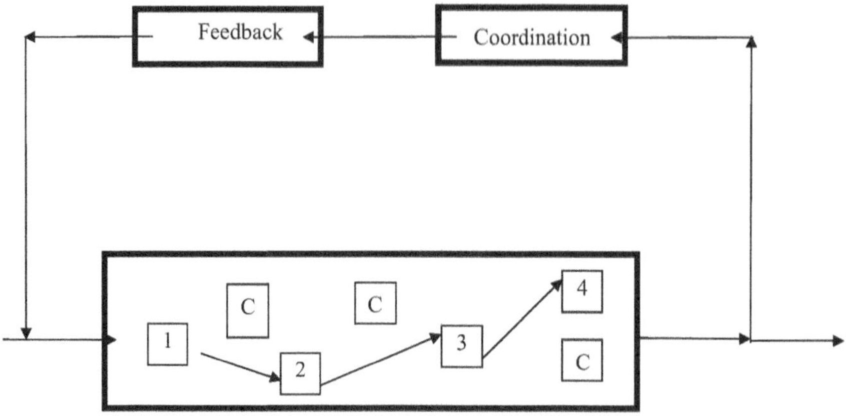

Fig. 4.13 Operational mechanism for a factory

Industry 4.0. It is an application of the Industrial Internet built with sets of hardware and software that collectively enable processes to govern themselves through machine learning and cognitive computing [25, 27].

The result is a level of operations that was previously impossible to achieve.

Downtime is predicted and prevented, waste and defects are reduced, surplus production is minimized, machine behavior is optimized as conditions change, and systems can make context-based next best actions. Connected devices in the factory report their status, giving operations personnel and decision-makers access to real-time, actionable information. Wearable technology tracks employee location and status in case of emergency. A global ecosystem of partners ensures that specific parts are replenished based on automated, real-time needs analysis. The list of applications increases.

But significant challenges remain in assessing, implementing, and operating Industrial Internet systems. From the perspective of control engineering, the smart factory can be viewed as a closed loop system, as shown in Fig. 4.13 [25].

In the center of the control loop is the network of equipments denoted in Fig. 4.13 by numbers. The equipments have cyber capabilities denoted by C. They have autonomy and collective behavior. Autonomy means that the equipment makes decisions by itself and no other entities can directly control its behavior. Collective behavior means that the equipments understand a common set of knowledge and follow a common set of rules for cooperation. Therefore, a society of equipments can yield a highly flexible manufacturing system, that is, a self-organized and self-reconfigurable system that may be considered as smart or intelligent. Through collaboration, the equipments try to align their behaviors to approach a system-wide goal.

The system performance is generally not optimal. This is because the equipments make decisions based on local information. As to manufacturing, load may not be balanced, efficiency may not be the highest, and deadlocks may arrive. The big data analytics block in the feedback channel serves to solve this problem. The machines communicate their state and process information to the block, and the distributed sensors transfer their sensed data to the block as well. Therefore, the global state of the system can be extracted from the real time system information.

Based on the computing ability, the block processes this big data in time to serve to coordinate the behaviors of the distributed smart equipments and contains feedback performance indictors to the self-organized network.

By this global optimization, the smart equipments are affected so that higher performance can be expected.

Figure 4.14 illustrates the difference between the traditional production line (Fig. 4.14a) and the smart or intelligent factory production system (Fig. 4.14b) [25].

The traditional production line is designed to produce the single type of products.

It generally consists of several machines and a conveyor belt. A single route for the product $1 \rightarrow 2 \rightarrow 3 \rightarrow 4$ is possible. The conveyor belt is not closed, that is, one end serves as the input and the other end serves as the output, and the machines are deployed along the line. When the unfinished products flow through the line from the input to the output, each machine performs its predetermined part of task. Generally, no redundant machines exist, and the conveyor belt is carefully tailored. The machine has its own independent controller, but the communication between machines is uncommon. The production line is basically linear.

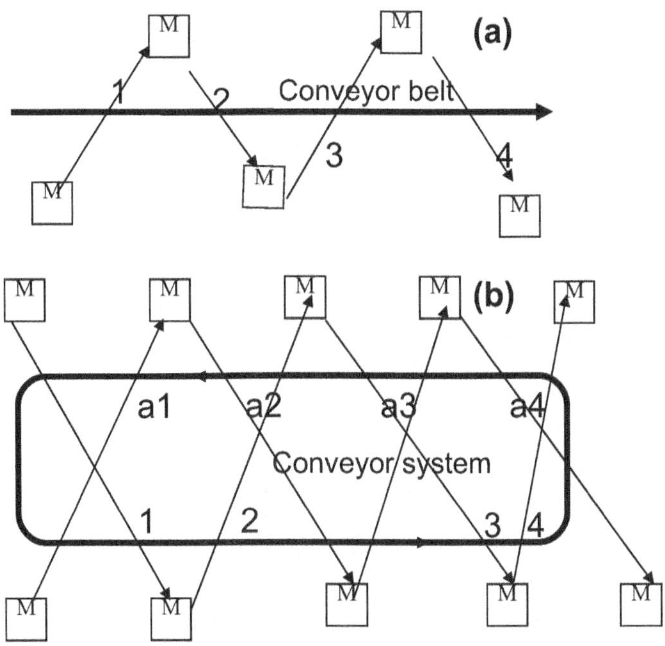

Fig. 4.14 Traditional production line and smart production system

Table 4.1 Comparison between traditional and smart factory

Nr	Traditional production line	Smart production system
1	Limited fixed resources	Diverse variable resources
2	Fixed routing	Dynamic routing
3	Shop floor central control	Comprehensive connections
4	Separated layer divergence	Deep convergence
5	Independent control	Self-organization
6	Isolated information	Big data
7	Simulation model	Digital twin

By contrast, the smart factory production system aims to process multiple types of products. Multiple routes for example 1 → 2 → 3 → 4 or a1 → a2 → a3 → a4 should be considered. Thus, from the view of the single product type, the machines are redundant. The machines relay on negotiation to reconfigure themselves to adapt to product type variation. The conveyor belt should be closed to support various production routes, so there is not definite input or output. The smart conveyor become a cyber-physical system and need a digital twin for efficient modeling and simulation [23].

Described as consisting of a physical entity, a virtual counterpart, and the data connections in between, the digital twin is explored as a mean of improving the performance of physical entities through the virtual counterpart.

These fundamental differences deliver some key features for the smart factory.

The smart production system couples the linear and cyclic capabilities.

The traditional and smart production systems coexist in duality.

In terms of dual systems theory, the features recorded in first column in Table 4.1 are more close to System 1 while the last column aspects are more close to the System 2 [8].

Conventional manufacturing lines are synchronous, with predefined workflows based on production work orders running in enterprise business systems. Production steps are centrally communicated to each manufacturing station synchronized with the assembly line. In contrast, smart factory is based on asynchronous manufacturing, with components in the production flow using self-identification technology to inform each machine and operator what needs to be done to produce the customized end product at each step of the production process. The use of flexible machines that adapt to the requirements for the part being made is another capability of smart industry. This achieves a highly flexible, lean and agile production process enabling a variety of different products to be produced in the same production facility. Mass customization allows the production of small lots due to the ability to rapidly configure machines to adapt to customer-supplied specifications and additive manufacturing.

A developed organizational mechanism is shown if Fig. 4.15.

The developed framework uses some elements of the cyber-physical-social systems for command and control [14].

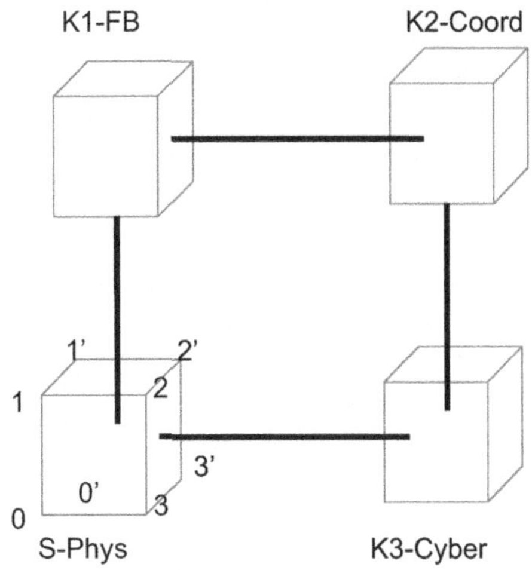

Fig. 4.15 Developed operational mechanism

The main levels or stages of organization and the corresponding abbreviations are:

- S-Physical-Phys
- K1-Feedback-FB
- K2-Coordination-Coord
- K3-Cyber-Cyber.

Figure 4.15 schema generalizes the schema shown in Fig. 4.13 replacing its elements by networks.

S corresponds to equipments networks, K1 to cognitive network, K2 to organizational coordination network and K3 to the cyber network.

Details or sub-levels may be considered at any of these four main levels.

These sub-levels or sub-stages have been introduced in development stage theory, DST, by Fischer [9]. They have been denoted 0-singles, 1-mappings, 2-systems, 3-SoS. Obviously we need to find interpretations of sub-levels for the particular domain considered as isomorphic to DST.

For instance, for the level S we may observe the sub-levels [14]:

0-devices, 1-sensors, 2-enablers, 3-communications.

The sub-levels 0', 1', 2', 3' denotes modifications or improvements.

This means that a conservative development strategy is selected for sub-levels.

Figure 4.16 shows the polytope 6D for developed operational mechanism including modifications and improvements.

A conservative development strategy is selected for the main stages or levels.

Phys', FB', Coord', Cyber' denotes the modified main stages.

Modification may be operated at the stage and at the sub-stage echelon.

Multiplicity of scales is also significant for conveyors.

Figure 4.17 shows Polytopic Roadmaps for conveyors evolution from 2D to 4D.

Figure 4.18 shows a conveyor 5D.

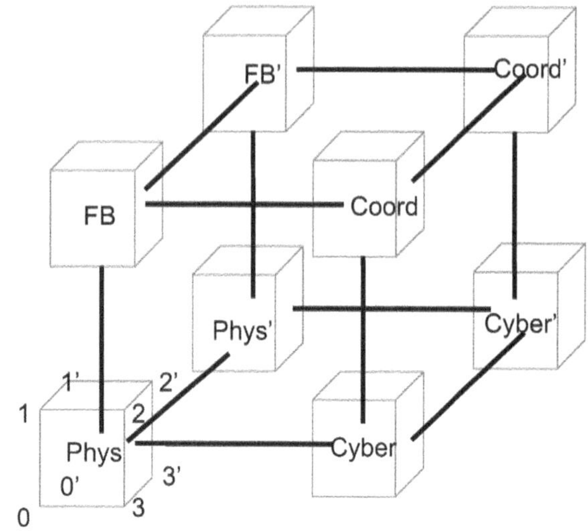

Fig. 4.16 Polytope 6D for developed operational mechanism

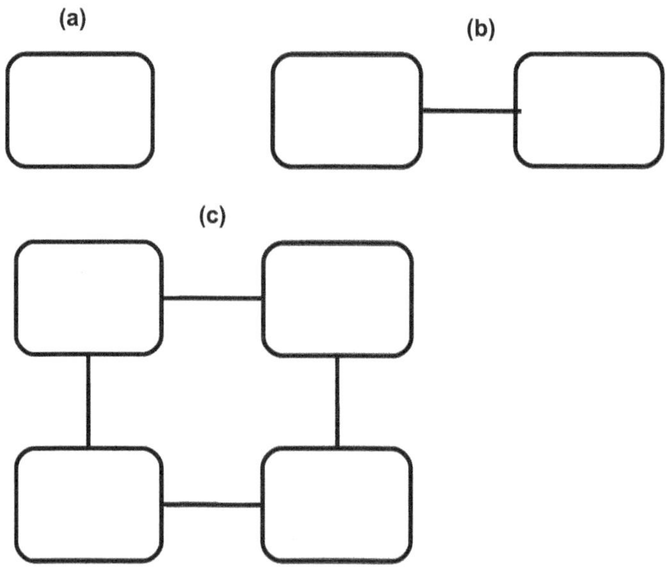

Fig. 4.17 Polytopic Roadmaps: conveyors 2D to 4D

The four lines connecting the two 2D rounded rectangles are represented in Fig. 4.17b by a thick hyper-line. In this way the 3D frame allows a simplified representation. All dimensions higher than two may be represented using thick hyper-lines in order to visually simplify the resulting structures. Using the thick hyper-line the 3D frame is shown in Fig. 4.17b as a "line of rounded rectangles". The dimension is $1 + 2 = 3$.

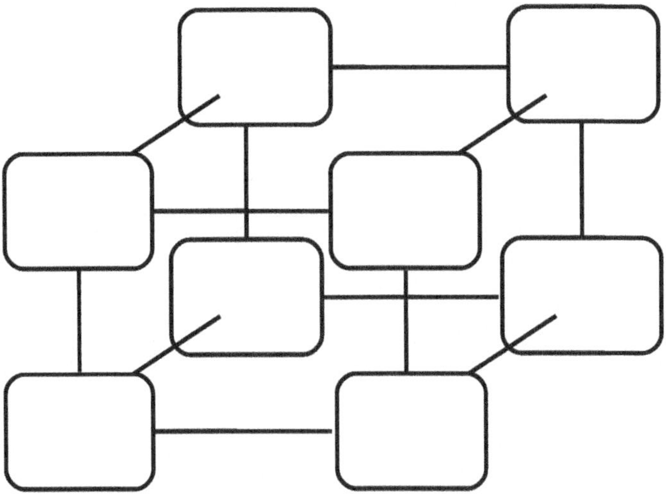

Fig. 4.18 Conveyor 5D

Figure 4.17c shows a 4D rounded rectangle. The internal rectangles again should have the corresponding rounded corners connected as for the 3D rounded rectangle. Using the thick hyper-lines, the 4D rounded rectangles is shown as "rounded rectangle of rounded rectangles". The dimension is $2 + 2 = 4$.

Figure 4.18 shows a 5D conveyor. Using the thick hyper-lines, the conveyor is shown as "cube of rounded rectangles". The dimension is $3 + 2 = 5$.

Figure 4.19 shows a Periodic Table for conveyors.

The columns correspond in succession to: dots, lines, conveyors (rounded rectangles), sets of parallel conveyors (cubes).

Dots correspond to fixed point of assembly.

The first row of the table starts from dots and finishes with a 3D-cube. Two or more parallel conveyors may be grouped in a cube.

The second row of the table starts from cubes (3D) continues with line relating two 3D-cubes (4D-cube) and finishes with cube of cubes represented as 6D-cubes (cube of cubes). For this the dimension is $3 + 3 = 6$.

The third row of the table starts from 6D-cubes continues with line relating two 6D-cubes (7D-cube) and finish with 9D-cubes (cube of cubes of cubes). For this the dimension is $3 + 3 + 3 = 9$.

The next row not included in the table will correspond to 9D, 10D, 11D and 12D conveyors.

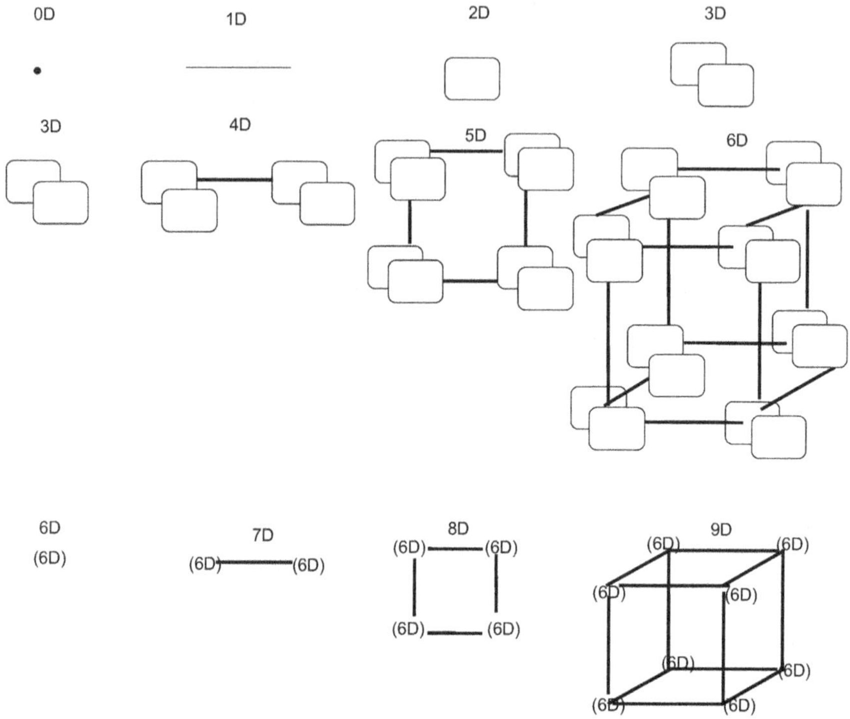

Fig. 4.19 Periodic Table for conveyors

4.4 Assembly

A new standard, Hermes, was created to serve as a backbone for board-flow data management in smart electronics manufacturing facilities, by providing an independent method of machine-to-machine (M2M) communication for surface mount technology, (SMT), assembly.

Taking advantage of modern transmission control protocol communication and Extensible Markup Language, XML data format technologies, Hermes is effectively a replacement of the SMEMA (Surface Mount Equipment Manufacturers Association) standard (IPC-SMEMA), which is obsolete. IPC denotes Institute of Printed Circuits.

The Hermes standard was initiated by equipment companies and was launched in 2017 as an open standard under the auspices of the Hermes Standard Initiative. It has subsequently been recognized by IPC under the designation IPC-Hermes.

All suppliers of SMT equipment are free to join the Hermes Standard Initiative and to participate and it counts roughly more than 50 companies amongst its membership.

Based on Transmission Control Protocol, TCP, and XML, the Hermes standard uses modern communication technology and standardized data formats for M2M communication. Its purpose is to transmit barcode information and board related data such as dimensions, board identifiers and product type identifiers together with the board along the machines of different vendors in an SMT line.

This means that a maximum of only one barcode scanner is required in the line. Besides this, the product itself becomes the driver of changes along the SMT line, so that manual adjustment of conveyor belt widths becomes unnecessary. Likewise, the information provided with a board can be used to automatically change the program according to the product type.

There are a number of clear benefits this offers to manufacturers of electronic assemblies. First of all, if all the machines on a production line are talking the same language, regardless of who made them, closed-loop handover is more easily achievable.

Since it is protocol based rather than signal based, it is easier to adjust and expand for integrating further information. It also uses standard components in the form of Ethernet cables, plugs and interfaces that are inexpensive and readily available. If supported by the equipment, cables can even be done away with in favor of standard Wi-Fi.

New equipment can be used alongside existing SMEMA-only devices if required, but the full benefit will only unfold in lines where the majority of machines are compliant with IPC-Hermes.

The Hermes Standard Initiative is open to all equipment suppliers who want to actively participate in bringing the benefits of Industry 4.0 to the customers.

Figure 4.20 illustrate the Transmission Process Protocol, TCP connections in an assembly line.

Figure 4.21 illustrate the communication sequence for board transport.

Figure 4.22 shows the polytope 4D associated to the assembly line.

A conservative development strategy is selected.

The direct sequence S → K1 → K2 → K3 corresponds to the upstream assembly stages.

The reverse sequence K3′ → K2′ → K1′ → S′ corresponds to the modified downstream assembly or disassembly stages.

Figure 4.23 illustrates the duality between assembly (Fig. 4.23a) and disassembly (Fig. 4.23b).

Figure 4.23a shows the U graph while Fig. 4.23b shows the D graph.

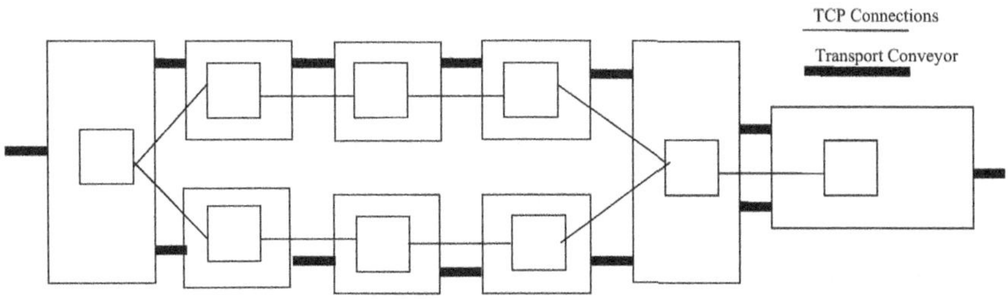

Fig. 4.20 TCP connections in a line

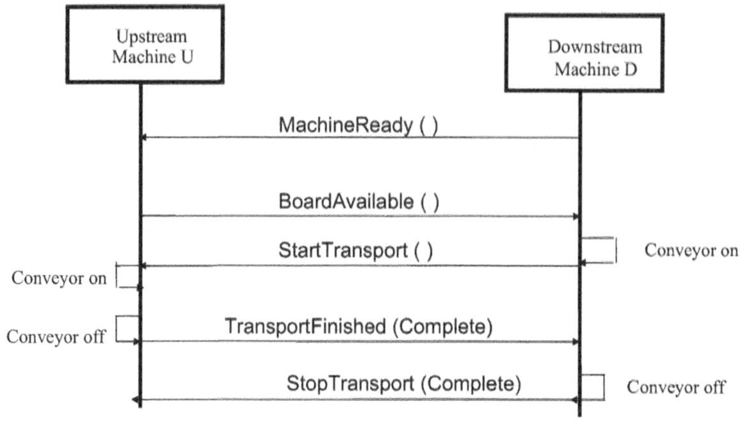

Fig. 4.21 Communication sequence for board transport

Fig. 4.22 Polytope 4D for assembly line

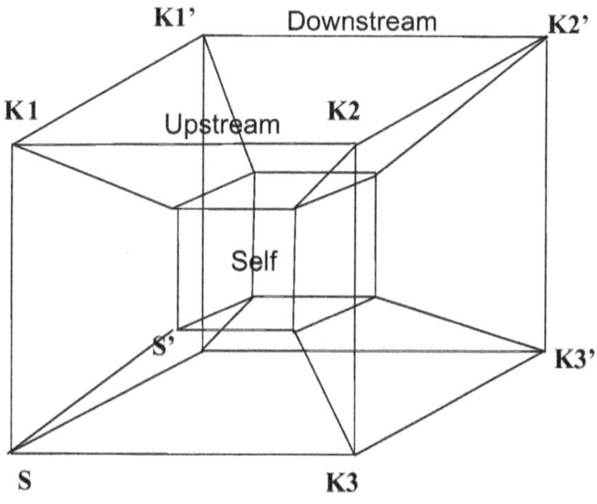

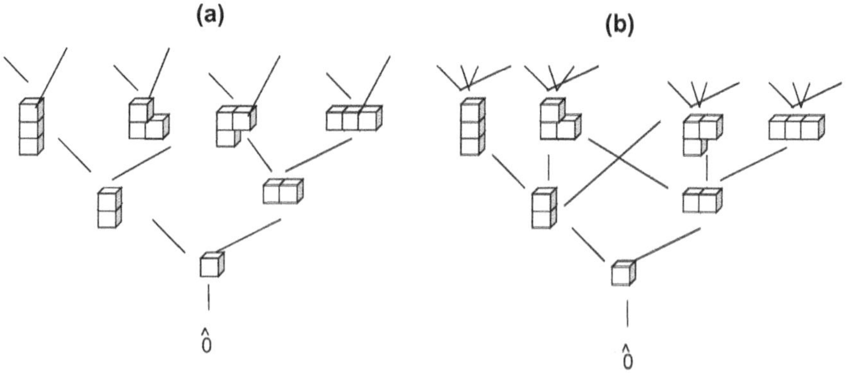

Fig. 4.23 Dual assembly and disassembly

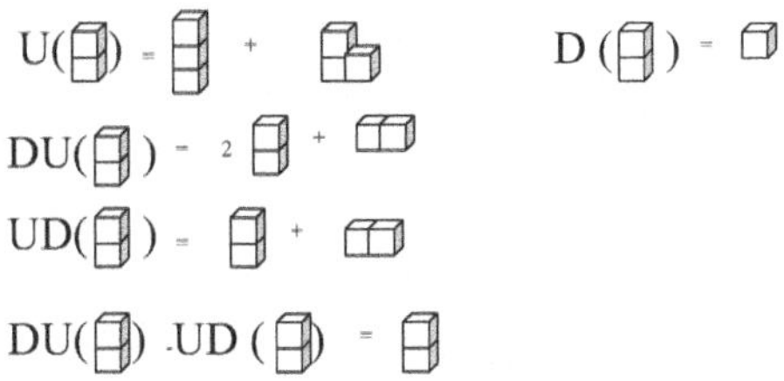

Fig. 4.24 Commutation conditions for dual assembly and disassembly

The U-tree describes the assembly while the D-tree describes the disassembly. It is the situation encountered in the study of dual graded graphs.

Figure 4.24 illustrates the commutation conditions for dual assemblies.

A self-evolvable construction should be able to swing between dual assemblies.

Reconfigurable tableau or lattices may be built by cube-shape modular constructions Gilpin and Rus [10, 11] presented an innovative approach to rapidly configure active objects using a large collection of smart particles (sand, pebbles, tiles, cubes) capable of communicating and bonding with their neighbors. Traditional manufacturing create devices using a hierarchical approach that is organized as a sequence that includes mechanical design, fabrication, electronic design, boards production, assembly and finishing. In contrast, it is possible to package the necessary mechanical, perceptual and computational capabilities in small, smart building blocks that integrate actuation, sensing communication and control. Reconfiguration in the system proposed by Gilpin and Rus [10, 11] consists of selecting the correct aggregation of these modules for a required shape or design. This work builds on a body of research addressing programmable matter, modular self-reconfiguring systems and smart systems.

References

1. Ammon, J.: Hypercube Connectivity within ccNUMA Architectures. SGI Origin Team, Mountain View, CA (1998)
2. Commenge, J.-M., Saber, M., Falk, L.: Methodology for multi-scale design of isothermal laminar flow networks. Chem. Eng. J. **173**, 541–551 (2011)
3. Coppens, M.O.: Nature inspired chemical engineering learning from the fractal geometry of nature in sustainable chemical engineering. Fractal Geom. Appl.: A Jubilee Benoit Mandelbrot **72**, 07–32 (2004)
4. Coppens, M.O.: Scaling-up and-down in a nature-inspired way. Ind. Eng. Chem. Res. **44**(14), 5011–5019 (2005)
5. Coppens, M.O.: A nature-inspired approach to reactor and catalysis engineering. Curr. Opin. Chem. Eng. **1**(3), 281–289 (2012)
6. Ehrfeld, W.: Process intensification through microreaction technology. In: Stankiewicz, A., Moulijn, J.A. (eds.) Re-engineering the Chemical Process Plant-Process Intensification, pp. 155–175. Marcel Dekker, Inc., NY (2007)
7. Ehrfeld, W., Hessel, V., Lowe, H.: Microreactors: New Technology for Modern Chemistry, Wiley-VCH (2000)
8. Evans, J.S.B.T.: Dual-processing accounts of reasoning, judgment, and social cognition. Annu. Rev. Psychol. **59**, 255–278 (2008)

9. Fischer, K.W.: A theory of cognitive development: the control and construction of hierarchies of skills. Psychol. Rev. **87**, 477–531 (1980)
10. Gilpin, K., Rus, D.: Modular robot systems: from self-assembly to self-disassembly. IEEE Robot. Autom. Mag. **17**(3), 38–53 (2010)
11. Gilpin, K., Rus, D.: A distributed algorithm for 2D shape duplication with smart pebble robots. In: Robotics and Automation (ICRA), IEEE International Conference, pp. 3285–3292, IEEE (2012)
12. Jones, D., Snider, C., Nassehi, A., Yon, J., Hicks, B.: Characterising the digital twin: a systematic literature review. CIRP J. Manuf. Sci. Technol. (2020)
13. Kashid, M.N., Gupta, A., Renken, A., Kiwi-, L.: Numbering-up and mass transfer studies of liquid–liquid two-phase microstructured reactors. Chem. Eng. J. **158**(2), 233–240 (2010)
14. Liu, Z., Yang, D.S., Wen, D., Zhang, W.M., Mao, W.: Cyber-physical-social systems for command and control. IEEE Intell. Syst. **26**(4), 92–96 (2011)
15. Luo, L. (ed.): Heat and Mass Transfer Intensification and Shape Optimization: A Multi-scale Approach. Springer Science & Business Media (2013)
16. Pistoresi, C., Fan, Y., Luo, L.: Numerical study on the improvement of flow distribution uniformity among parallel mini-channels. Chem. Eng. Process. **95**, 63–71 (2015)
17. Ruthven, D.M.: Principles of Adsorption and Adsorption Processes. Wiley, New York (1984)
18. Saber, M., Commenge, J.M., Falk, L.: Rapid design of channel multi-scale networks with minimum flow maldistribution. Chem. Eng. Process. **48**, 723–733 (2009)
19. Saber, M., Commenge, J.-M., Falk, L.: Microreactor numbering-up in multi-scale networks for industrial-scale applications: impact of flow maldistribution on the reactor performances. Chem. Eng. Sci. **65**, 372–379 (2010)
20. Seitz, C.L.: The cosmic cube. Commun. ACM **28**(1), 22–33 (1985)
21. Stankiewicz, A., Moulijn, J.A.: Process intensification. Ind. Eng. Chem. Res. **41**(8), 1920–1924 (2002)
22. Stankiewicz, A., Moulijn, J.A. (eds.): Re-engineering the Chemical Process Plant-Process Intensification, pp. 155–175. Marcel Dekker, Inc., NY (2007)
23. Tao, F., Qi, Q., Wang, L., Nee, A.Y.C.: Digital twins and cyber–physical systems toward smart manufacturing and industry 4.0: correlation and comparison. Engineering **5**(4), 653–661 (2019)
24. Thiel, T.: The design of the connection machine. Des. Issues **10**(1), 5–18 (1994)
25. Wang, S., Wan, J., Li, D., Zhang, C.: Implementing smart factory of industrie 4.0: an outlook. Int. J. Distrib. Sens. Netw. **12**(1), 3159805 (2016)
26. Yang, R.T.: Gas Separation by Adsorption Processes. Butherworths, Boston, MA (1987)
27. Zuehlke, D.: Smart factory–from vision to reality in factory technologies. IFAC Proc. **41**(2), 14101–21408 (2008)

Chapter 5
Technologies

5.1 Flow-Sheets

Flow-sheet structure is the product of individual adaptation to particular input and environmental demands. As a result, flow-sheet structures are not static properties of single installations, but instead are dynamic products of installation-in-context. A change in situational demand, input domain or environmental context necessitates a change in the structure and content of specific actions and technology.

Flow-sheet generation and analysis is a preliminary stage of process design, in which the building blocks of the process are chosen, their interconnections are specified, and the results evaluated. Separation device are often the largest and heaviest items of process equipment in a production system. Consequently, it is of interest to improve its design. Studies identify four approaches for solving the combinatorial problem of synthesis of separation sequences, namely: evolutionary, algorithmic, heuristic and artificial intelligence, AI, methods. In evolutionary methods a processing system is devised, analyzed and changed in one or more ways so as to improve it. Evolutionary strategy includes the following steps: generate an initial process, define the evolutionary rules, and determine the evolutionary strategy. Algorithmic methods transform the synthesis step into large-scale optimization problems and different optimization strategies developed in the field of dynamic programming are used to solve it. Because of the large combinatorial nature of the problem, for complex flow-sheets it is unsuitable to investigate thoroughly all possible sequences. Simple heuristic and qualitative rules are generally used to generate or to select initial sequences which are likely to be nearly optimal and which deserve a more detailed investigation.

An approach leading to a simple computerized algorithm able to select from multiple separation techniques was proposed and applied to the situation when a full separation of the components is required [13, 14].

A set of properties and heuristics related to the separation process is selected and ordered in a hierarchy. For example in the case of distillation the properties

O. Iordache, *Polytopic Roadmaps*, Studies in Systems, Decision and Control 368 ,
https://doi.org/10.1007/978-3-030-75630-7_5

could be: volatility, feed composition, aromatic character, and so forth If different separation techniques are possible properties as: volatility, the dipole moment, the normal freezing point, the maximum molecular diameter etc., are accounted for [8]. It is considered that the separation must be performed on account of these properties ranged in hierarchical order of importance.

The first step in quantifying the concept of similarity for chemical species in a mixture is to list the most important properties of such species. To every species can be associated a vector the components of which take only two values "1" or "0" where "1" means the presence of a given property whereas "0" means its absence. Otherwise "1" may correspond to a high value of the property, for example high volatility, whereas "0" corresponds to a low value. Thus the vectors associated to the species are:

$$\underline{i} = \langle i_1, i_2, \ldots, i_k, \ldots \rangle; \underline{j} = \langle j_1, j_2, \ldots, j_k, \ldots \rangle$$

where i_k, j_k are either "1" or "0". Binary characterization according to the presence ("1") or to the absence ("0") of a given property was used. The use of the real properties instead of Boolean ones has been also tested. A hierarchy of the properties and heuristics is required. For instance, it is considered that the property indexed by i_1 is more significant that the property indexed by i_2 and so on in the order of the coordinates in the associated vectors. The initial choice of this hierarchy introduces an important subjective character into the analysis where precisely the expertise of the analyst plays an essential part.

Consider as an example the separation of a group of four closely-boiling compounds (xylene isomers).

The corresponding vectors of properties have been presented in Table 5.1.

The properties considered in Table 5.1 are: volatility (relative to o-xylene), the dipole moment, in Debye D, the normal freezing point, in degrees Kelvin K, the maximum molecular diameter, in angstroms Å.

Table 5.2 contains in Boolean form the first three columns shown in Table 5.1.

Here low "0" and large "1" values have been classified relative to the mean values (Table 5.2).

Tables as 5.2 may be considered as context tables.

Table 5.1 Input information-separations of xylene isomers

Data	Name	Volatility	Dipole, D	Freezing point, K	Maximum Diam, Å
1	m-xylene	1.6	0.4	225.4	8.33
2	o-xylene	1	0.6	248.1	7.8
3	p-xylene	1.63	0.0	286.6	8.67
4	Ethyl benzene	1.73	0.58	178.4	9.00

The units are: D—Debye, K—Kelvin degrees, Å—Angstrom

Table 5.2 Boolean input information-separations of xylene isomers

Data	Name	Volatility	Dipole	Freezing point	Diameter
1	m-xylene	1	1	0	0
2	o-xylene	0	1	1	0
3	p-xylene	1	0	1	1
4	Ethyl benzene	1	1	0	1

For such context tables we can apply FCA. FCA originates conceptual lattices and these in turn technologies.

We replace classification based technologies as discussed by Iordache et al. [14] by technologies based on FCA applied as a classification tool.

Usually we can select between several technologies.

This is the situation shown in Figs. 5.1 and 5.2.

Figure 5.1 shows the contexts Polytopic Roadmaps from 1D to 5D.

Figure 5.1a corresponds to the separation based on a unique property.

Figure 5.1b shows a typical context. Rows correspond to objects as example mixtures to be separated. Columns correspond to properties or attribute allowing separations.

Figure 5.1c shows several contexts. Any table corresponds to a specified technological line.

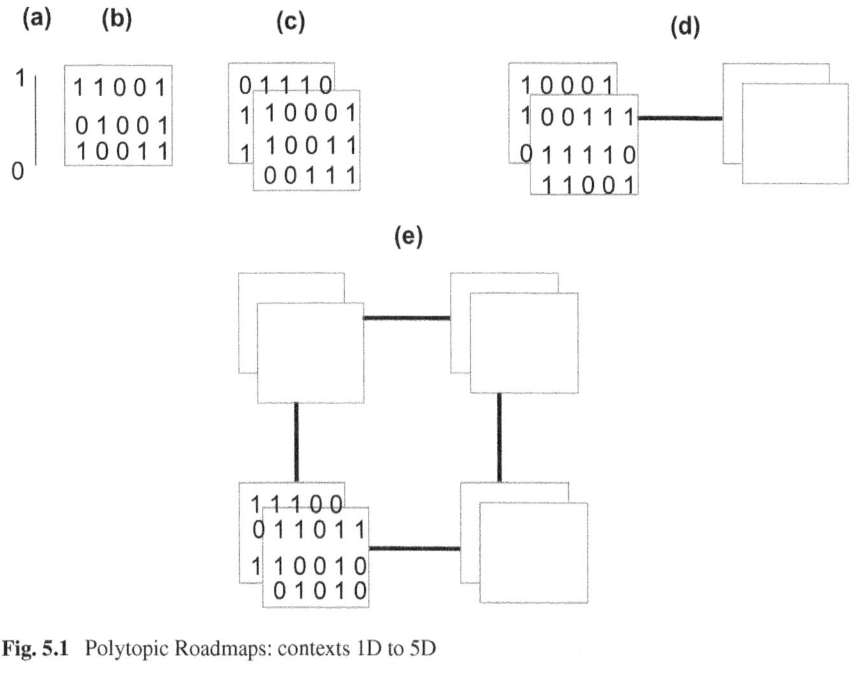

Fig. 5.1 Polytopic Roadmaps: contexts 1D to 5D

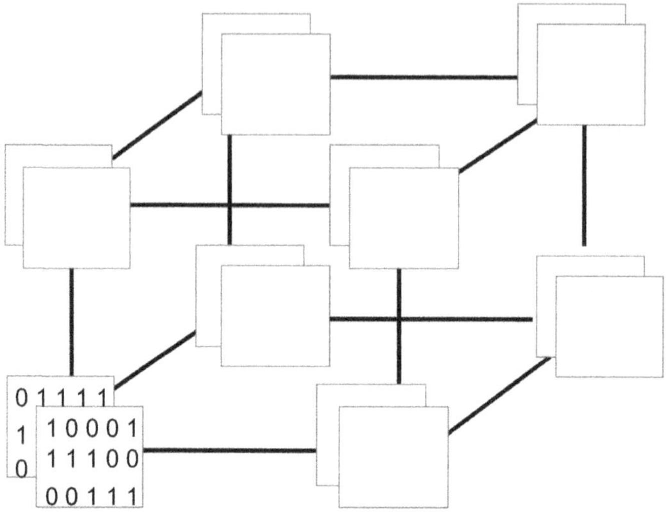

Fig. 5.2 Contexts 6D

Figure 5.1d shows a new dimension for instance the plant.

Plants have several technological lines.

Figure 5.1e shows a new dimension for instance the country operating the plant.

Figure 5.2 shows the contexts 6D.

Figure 5.2 outlines a new dimension for instance the continent operating the plant.

Observe that the hierarchy of spatial scales imposes the hierarchy of scales for technologies.

Figure 5.3 shows a Periodic Table for formal contexts. A formal context can be represented by a table specifying which objects fall under which properties. This suggests that any formal context may be associated to classification purposes and to technologies.

The columns correspond in succession to: one property or attribute (lines), objects-properties tables (rectangles) that are typical contexts, and several tables of object-properties (cubes), that are sets of parallel contexts.

The first row starts from one line and finish with a 3D-cube.

The second row starts from line relating two cubes (4D) and finish with cube of cubes represented as 6D-cubes.

The third row starts from line relating two 6D-cubes (7D) and finish with 9D-cubes (cube of cubes of cubes).

The next row of the Periodic Table, unincluded in this figure will show 10D, 11D and 12D frames.

Column or rows in the context tables may interact according to the rules established by the tick-lines. This ensures the flow-sheeting flexibility and self-evolvability.

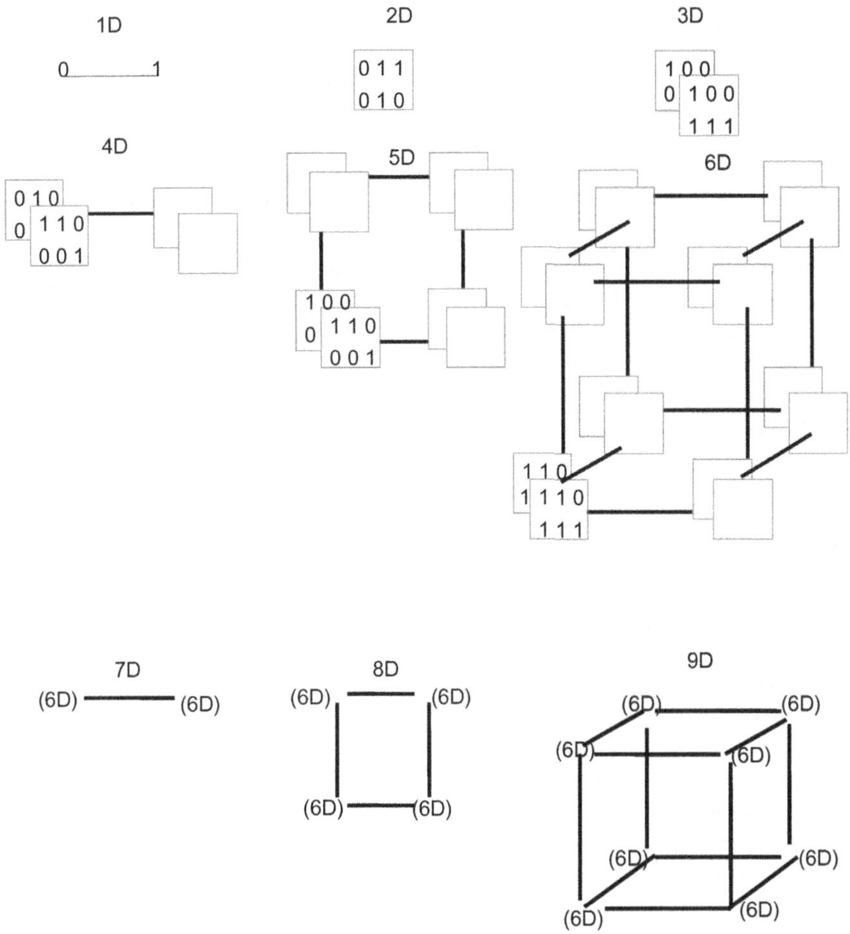

Fig. 5.3 Periodic table for contexts

5.2 OLAP

OLAP is an architecture used to provide users with the ability to perform dynamic data analysis. By using OLAP tools, users gain access to the analytical database. This type of access provides decision makers with the potential to improve the understanding of changes, the ability to identify or generate possible solutions for a variety of decision problems, and the ability to develop tactical and strategic formulae regarding variables.

An important additional rule in OLAP is the incremental database refresh. This enables the database to be updated only in the area where changes have taken place.

At an application level, OLAP is product-specific, and varies by implementation. From the study of OLAP, one concludes that multidimensional viewing would perform best if the data were stored in multi-dimensional data bases. These have performance advantages when the data is dense. Since the introduction of OLAP, many users have, argued in favor of relational-OLAP, (ROLAP), that is, OLAP tools based on relational database technology.

ROLAP systems have a relational based analytical database.

This is a database suitable for use for decision-making purposes. A significant characteristic of the analytical database is its incorporation of time within its data structures.

Multiple dimensions viewing, is a key OLAP objective [15].

A dimension is a data category for instance, product, district, time. This category may have one or more different characteristics which are the dimension values, for instance, products may be a, b, or c, district may be north, east, west, or south and time may correspond to different years. A data value in a multidimensional table then reflects a particular business perspective for instance., sales for a corresponding combination of dimension values, or sets of dimension values for instance, sales for product a, district south, for recent years.

In relational database terminology, dimensions correspond to the attributes of a relational table, while the dimension values correspond to the attribute's domain or the set of possible attribute values. The multidimensional data table values correspond to the records of a relational table. Multidimensional data tables reflect a specific view or perspective on the data for instance sales by region, and product, over time. The benefit is that the data fits the user's perspective. One example to visualize this multidimensional viewing is a worksheet.

Figure 5.4 shows Polytopic Roadmaps: OLAP from 2D to 5D.

A single worksheet is two-dimensional that is, one dimension being the columns, the other dimension being the rows. A stack of worksheets is three-dimensional. This structure is known as a cube. A stack of cubes viewed along a new dimension is considered a four-dimensional hypercube. Along a further dimension is five-dimensional, and so on for all dimensions employed in a multidimensional table.

For example, suppose an analyst wishes to analyze data concerning some products, over some period of time, for some salesman from some particular sales sites.

In dimensional terms, the data dimensions are product, time, salesman and site.

A two-dimensional view of time and product is seen in Fig. 5.4a. Here y1 and y2 denote the years, while a, b, c denotes the product. In FCA terms objects are time dimension while products are attributes.

Considering the third dimension, for instance, site, produces a stack of time–period tables one for each site, as in Fig. 5.4b. A, B, C may correspond to sites. The addition of another dimension in this case, salesman, along one axis, creates a four-dimensional frame (Fig. 5.4c). It is possible to create a five-dimensional view by adding another dimension along another axis. Nevertheless, it is difficult to visualize more than four dimensions at a time.

These data-views are user-reconfigurable. Dimensions are not statically set in columns or rows, or along one axis or another. To use the OLAP terminology, the user

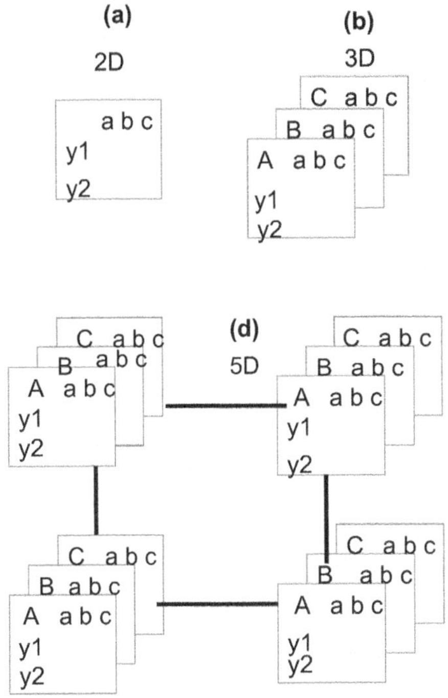

Fig. 5.4 Polytopic Roadmaps: OLAP 2D to 5D

has the ability to roll-up, drill-down, slice, dice or nest his data. We classify these functions as multidimensional data viewing and as multidimensional data manipulation functions. The viewing functions then are slice, dice and nest.

The manipulation functions are drill-down and roll-up.

Slice is any two-dimensional slice of the data. Slice is the dropping of the dimension being sliced.

Dice is the rotation of the hypercube to reveal another, different slice of data that is, the revealing of data for a different set of dimension values.

Nesting is the ability to nest dimensions that is, display values from one dimension within another dimension.

Multidimensional viewing has received attention because of its close relationship with OLAP and its direct relationship with databases. There are two main multidimensional data manipulation functions: drill-down and roll-up.

Figure 5.5 shows the 6D framework of OLAP for growing complexity.

Figure 5.6 shows a Periodic Table for OLAP.

The columns correspond in succession to: lines, years-product tables (rectangles), and several sits, denoted A, B of year-product tables grouped as cubes.

The first row starts from one line and finish with a 3D-cube.

The second row starts from line relating two 3D-cubes (4D-cube) and finish with cube of cubes represented as 6D-cubes.

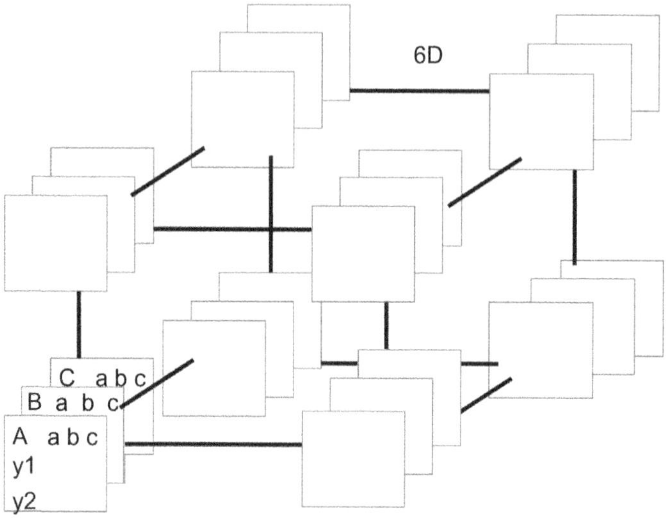

Fig. 5.5 OLAP 6D

The third row starts from line relating two 6D-cubes (7D-cube) and finish with 9D-cubes (cube of cubes of cubes).

The next row unincluded in this figure will show 10D, 11D and 12D frames.

Here y1 and y2 denotes the years, while a, b, c denotes the product.

A, B may correspond to the site.

5.3 Engineering Design

Companies are looking for models that they can use to improve organizational efficiency. Early models decomposed organizations into workers and management and ascribed roles to each. Newer models concern themselves with integration of the parts into the context of the outside world.

There exists an isomorphism between decomposing processes into unit operations which must then be integrated into total processes within the world market, and the process design activities. Chemical engineering concepts such as process control and information transfer appear to be applicable to organizational management as well.

The main objective of our study is to outline and implement the isomorphism between the cognitive-intelligent development stages, and the smart operations, equipments, technology and industry development stages. It is the general isomorphism with respect to the higher dimension development and cyclic control mechanisms.

After examining management processes and how information flows in a company, engineers need to explore different technologies for information management and

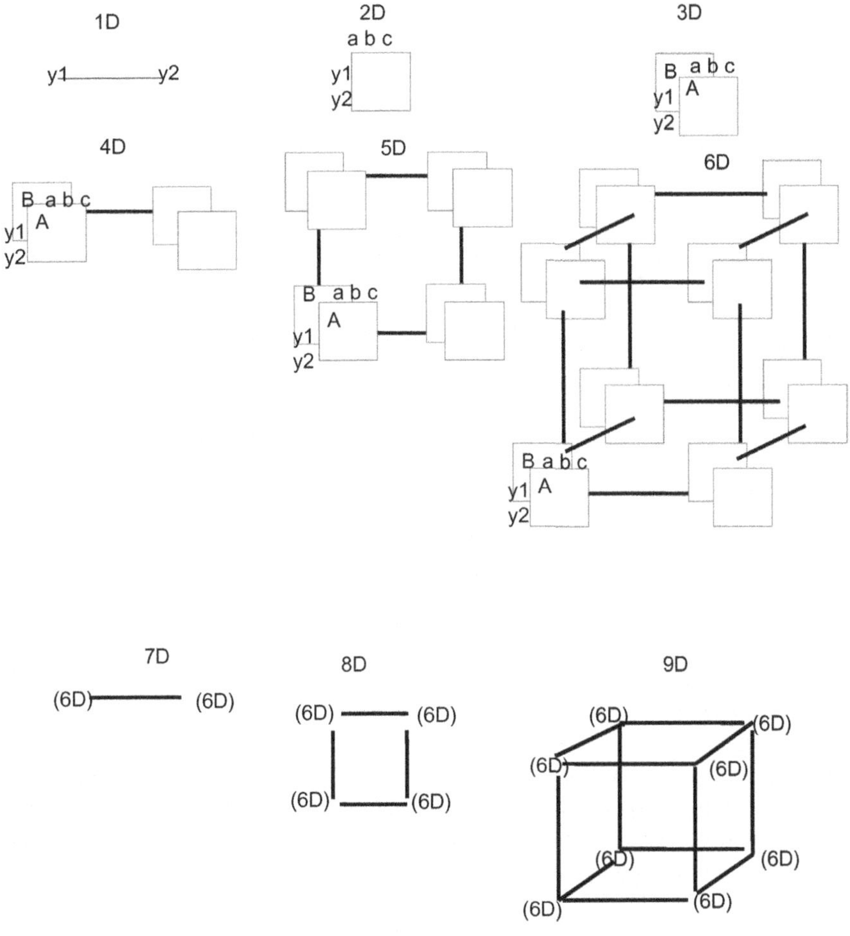

Fig. 5.6 Periodic Table for OLAP

consolidation, noting how well each permits the gathering, structuring and using of the data both to operate a company and to understand and improve its performance, particularly in its ability to carry out a complex process such as design [21].

Recent approach to organizational management analysis strongly parallels the state of chemical engineering process design analysis when process synthesis was emergent [11].

Institutional management may benefit from the systematic techniques developed for process synthesis. Furthermore, the needs of the process design community and the general business community in terms of information consolidation and management appear to be strongly similar or isomorphic [22].

It should have the same characteristics as a self-adaptive organism, adjusting to changes in the business environment and in technology. System prototypes, which

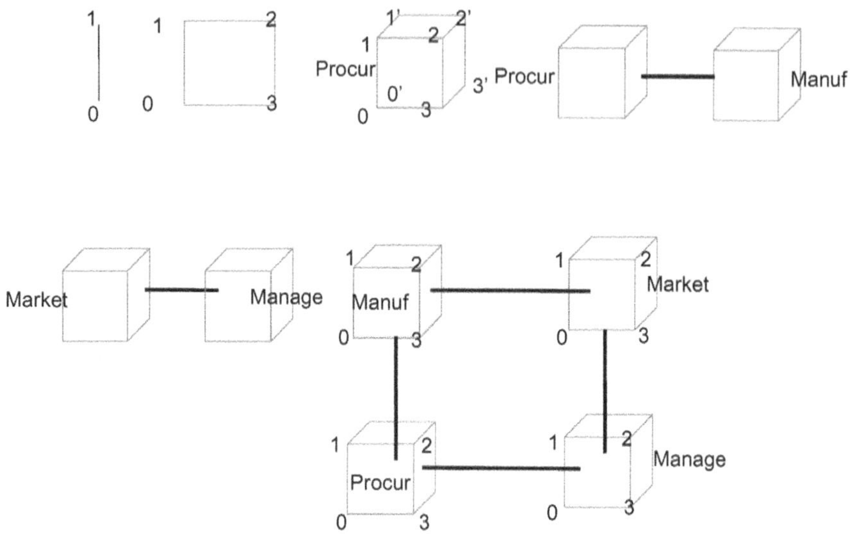

Fig. 5.7 Polytopic Roadmaps for engineering design

help understanding of business detail and dynamic complexity and which can be connected to form different business structures might be a useful starting point for business process synthesis. Furthermore, concepts in chemical engineering process control might be useful in the analysis of business environment and organizational dynamics. A basic difference is the human element in business processes. Participation with high performance teams, innovation, motivation to learn, and creativity are human characteristics for which information technology has little to offer at present.

However both business and process design success relies heavily on such characteristics.

Figure 5.7 shows Polytopic Roadmaps for engineering design.

The stages are: Procuring, Manufacturing, Marketing and Management.

The abbreviations are:

- Procuring-Procur
- Manufacturing-Manuf
- Marketing-Market
- Management-Manage.

While the design activity represents only a small fraction of total resources involved or total information flow in any corporation, it is strongly influenced by changes in information flow.

Figure 5.7 shows one perception of a total corporate information flow loop for a typical chemical company. All commercial businesses utilize work processes to convert some input to some output. Each of the stages represents a subdivision of the overall organization. In this example, the Procuring stage represents obtaining raw material or other inputs; Manufacturing represents the transformation process and

Marketing represents dialogue with the customer. Management makes up the final of the four major stages.

Further sub-stages depict major activities within each of the major divisions. It is likely that there are further sub divisions in most organizations.

The sub-stages of the main stages as shown in Fig. 5.7 have been studied by Bagajewicz [2]. A possible choice of sub-stages is as follows:

- Procuring
  - 0-environment, 1-supply analysis, 2-costing, 3-contracting
- Manufacturing
  - 0-construction, 1-design, 2-maintenance, 3-operations
- Marketing
  - 0-product analysis, 1-advertising, 2-customer service, 3-competitor analysis
- Managing
  - 0-human resources, 1-strategic planning, 2-information, 3-finance.

Almost everything done within the design sub-stage is influenced by some information from one or more of the other sub-stages,

Organizational control, as usually practiced, requires that information based on current data be approved by the management of each of the sub-stages before it is shared with other stages.

All information is not shared among all stages.

As information becomes more broadly shared and organizational culture changes, it is likely that the information available for design activities will also change.

Decisions that were made by parts of the organization and documented by providing functional or job specifications are more likely to be made by the design team. For example, data about costs of equipment, catalysts, and utilities, will be available in real time.

Figure 5.8 shows the 5D polytope of information flow loop for a typical chemical industry unit.

Figure 5.9 shows the polytope 6D for information flow loop including modifications and improvements.

A conservative development strategy was selected.

The direct sequence Procur → Manuf → Market → Manage corresponds to the main development stages. The reverse sequence Manage' → Market' → Manuf' → Procur' corresponds to the modified stages.

The modification may be operated at the stage or at the sub-stagelevel.

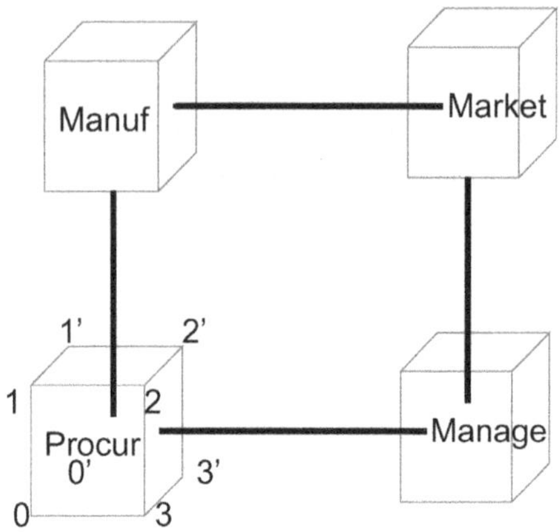

Fig. 5.8 Information flow loop for a typical chemical industry unit

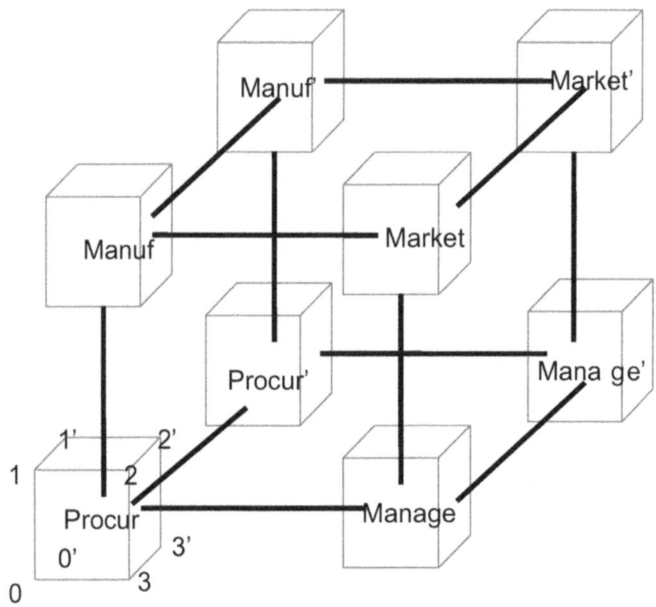

Fig. 5.9 Polytope 6D for information flow loop

5.4 Enterprise Control System

As the lifetime of many products that is, gets shorter, enterprises have to evolve their information system to account for changes in the internal and external business environment. Indeed, enterprises are using more and more software applications, such as the advanced planning and scheduling (APS) systems, the enterprise resource planning (ERP) system, the manufacturing enterprise system, (MES), and the supply

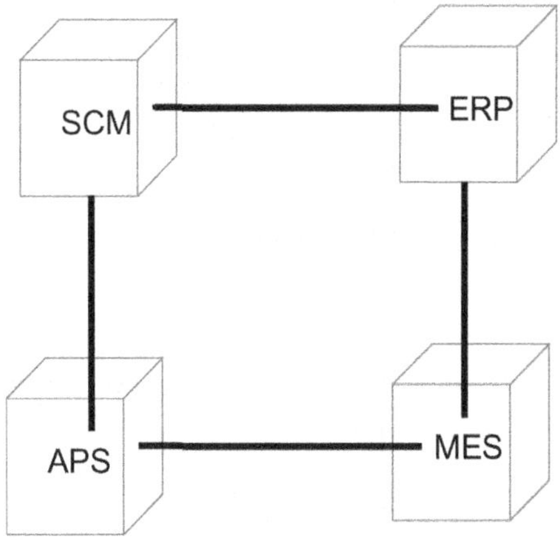

Fig. 5.10 Enterprise system: polytope 5D

chain management (SCM) system to deal with the increasing complexity of information flows. These applications are often distributed among facilities with various owners and many suppliers. This supply-chain environment is unstable, because consumers/customers have an increasing need to customize their products in ways that respond to the changes in supplier networks. That has come about as a result of the new phenomenon of extended enterprises. The business constraints cause issues with short-lived relationships about the way the systems interoperate. In a more general way, this happens not only to enterprise software applications but also to any enterprise system that handles information.

Figure 5.10 shows the architecture of an enterprise system as a 5D polytope.

The study of product-driven systems control has suggested a new paradigm: product-driven inter-operability [17].

Auzelle et al. [1] have studied interoperability issues and formalization, where enterprise systems are dealing with information management in a manufacturing context. In such context, a product embedding information about itself and being able to communicate with its environment may be qualified as an active product.

An active product defines the set of capabilities needed to produce it.

In order to be manufactured, those defined capabilities are provided by resources within a manufacturing system. These resources are controlled by a set of enterprise systems. When an active product joins this set of enterprise systems, we are demonstrating that the new system (existing enterprise system as well as the product) is characterized by specific properties derived from autonomy, belonging, connectivity, diversity, or emergence. From these new systems emerge new objectives.

These are the advanced planning and scheduling system, the enterprise resource planning system, the manufacturing enterprise system, and the supply chain management system.

This new system may then be assimilated to a system-of-systems by gathering all enterprise systems, considered themselves as autonomous systems, together with

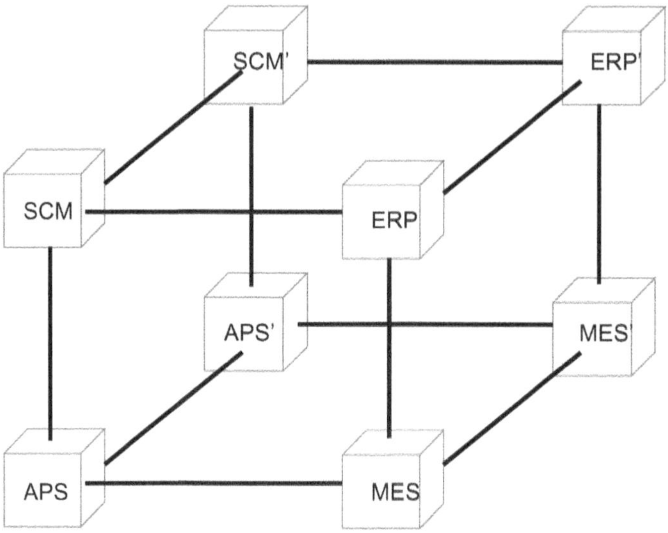

Fig. 5.11 Enterprise system: polytope 6D

the active product. All enterprise systems are also connected to the active product through an interoperating relationship that must be formalized.

A system-of-systems perspective for integrating enterprise-control system is required.

Figure 5.11 shows an enterprise system 6D polytope.

A conservative development strategy was selected.

The direct sequence APS → SCM → ERP → MES corresponds to the main development stages. The reverse sequence MES' → ERP' → SCM' → APS' corresponds to the modified stages.

In this system-of-systems-like context, the research addresses the formal mathematical definition of a product-driven interoperability relationship, taking into account the connectivity between an active product and a set of enterprise systems.

This relationship implies on the fly information exchange for ensuring on demand processes.

Technologies based on enterprise application integration (EAI) or service-oriented architecture (SOA) offer languages and protocols to aid exchange of service-based information. The research challenge is to explore the perspective of semantic relationships to give it more transparency and make interoperability seamless.

It is a need to explore systems engineering processes and standards, thus proposing a methodology to consider the formal semantics relationships between the involved enterprise system and the so-called active product. This approach may open new opportunities to manage the manufacturing process, considering each active product as a system that may interoperate with each enterprise system, enhancing then new planning perspectives in the manufacturing world.

5.5 Reality Inspired Engineering

Engineering deals with the processing of material, biological, cognitive and intelligent systems. Engineering also emphasize on the development of technologies used for the large scale production and on the manufacture of products with desired properties through specific processes. Interactions within and between processes are important as well as dynamics [9, 10].

Some models and knowledge in use in engineering have been established decades ago and have proven their performance to meet the twentieth century challenge of large-scale production at the lowest cost. Within the advent of high complexity and sustainable growth challenges now everywhere in the twenty-first century, priorities are renewed: renewable materials, energy efficiency, less waste, low emissions, and so on, but also small-scale production distributed in a region, close to consumers and human-aware production. All industries including chemical industry must follow the trace of ground-breaking industries as electronics, automatics, software and mechanics by implementing the elements of industry 4.0 of fourth industrial revolution and beyond these.

The request for regeneration concerns the operations, equipments, technologies, industries and includes calculus, design, simulation and control [9, 10].

Reality is supposed to be the source of inspiration on this way.

Reality is processing energy, matter, information, cognition and intelligence. It is also a model of structures and processes, whose performance, efficiency and resilience is challenged by engineering activity. Reality Inspired Engineering (RIE) is the research area of technology and engineering that seeks guidance from reality, using fundamental mechanisms to solve problems. According to Hartmann's ontology reality means:material-biological-cognitive-intelligence levels [12, 19].

These four levels or just part of them may serve as inspiration for engineering that is for artificiality.

The entrepreneurs have sought reality inspiration et al. times, but the reality-inspired approach is distinctive in that it seeks not only to understand, but also to implement, to impose, the fundamental mechanisms discovered to engineered systems, while explicitly accounting for the specific context and constraints of the application. This is inherently different from biomimetics or biomimicry, which has frequently been used in the past to develop new products and processes [16, 18].

Biomimetics is an interdisciplinary field in which principles from sciences as physics, chemistry and biology are applied to the synthesis of materials, synthetic systems or equipments that have functions that mimic biological processes.

Biomimetics is a bottom-up approach to innovation that seeks sustainable solutions to human challenges by emulating biology time-tested patterns and strategies. The goal is to create products, processes, and new ways of living that are adapted to life.

Biomimetics differs from the nature inspired engineering, NIE approach and from RIE approach in that it tends to presume that biology solution is the best without accounting for the different goals and context of natural and technical issues to address.

Proponents of biomimetics have been inspired by the old tradition were the natural laws ensures the best solutions. This old tradition supports a linear hierarchical model for development and for sciences.

Nature Inspired Engineering NIE is an approach focused on goals and contexts of natural and technical issues to address. It is more close to engineering and technological point of view than biomimetics that seems more close to scientific point of view [16].

Coppens [5, 6] has developed the field of NIE by enunciating four mechanisms that have different levels of engineering maturity:

- T1-hierarchical transport networks
- T2-force balancing
- T3-dynamic self-organization
- T4-control mechanisms in ecosystems, biological networks

NIE could help to improve performance, efficiency and resilience of engineering processes. To achieve this goal, some trends have been established for biology-inspired processes, spanning both micro- and large-scale processes.

Simply copying natural structures, tools or processes leads to sub-optimality because artificially made objects and processes do not operate in the same context or with the same goals as natural processes. Instead, one should decode the hidden mechanisms and dynamics so as to develop nature inspired engineered objects and processes.

By studying natural systems as for instance plants, cells or bacterial communities, from the perspective of an engineer, it is possible to draw indications to design better performing, more sustainable and efficient processes and materials for technological applications. Engineers seek guidance from nature as starting point in theoretically based, computationally aided, and experimentally validated designs.

The designs are optimized for criteria such as for example maximum chemical yield, selectivity and stability over time for catalytic systems. This approach was applied to catalysis and reactor engineering, energy and membrane separation technology [6].

At the heart of these nature-inspired engineering NIE designs lies a better fundamental understanding and control of physical or chemical phenomena across multiple length and time scales.

The effect of heterogeneity and confinement on diffusion and reaction phenomena was also studied by Coppens [4]. This study focus on patterns and symmetries those are frequent in nature, such as fractal structures, which link microscopic with macroscopic scales in an efficient and intrinsically scalable way.

The relation between different scales may be studied in terms of Polytopic Roadmaps with an emphasis on higher dimensions. In what follows the Reality Inspired Engineering, RIE procedures will be presented as a new perspective.

RIE and NIE are different since according to the acknowledged definitions Reality and Nature are different. RIE encompass NIE.

If we refer to sciences RIE is the general framework able to provide tools with which to distinguish and coordinate the various disciplinary and technological outcomes. Levels of reality encompass every science not only the sciences of Nature (as physics, chemistry biology and geology), but also cognitive and intelligence sciences. In other words reality is not limited to material or biological levels since cognitive (psychological) and intelligence (logico-mathematical, social) phenomena are real and reality too [19, 20].

The cognitive-intelligence development stages serving as inspiration for the engineering projects presented in this monograph are as real and reality as any tree or cell membrane or biological network.

RIE disagrees with the linear hierarchical model of biomimetics. Restriction to linear hierarchy appears from reality point of view as a partial perspective or an ontological error.

Reality includes cognitive and intelligence phenomena and these impose a cyclical or spiral model for development. If we refer to sciences we observe that the intelligence produces logics and mathematics, based on cognitive activity, this depends on biological phenomena, this in turn depends on material phenomena that in their turn are stated in logical and mathematical laws and with this we are back at the beginning of the cycle or spiral. Spiral results by coupling the linear and cyclic development.

The same cycle of spiral will be observed in the Polytopic Framework for RIE.

RIE is a basic method to increase dimensionality.

For RIE projects, the starting major stage is Reality.

Reality means four sub-stages or levels.

Based on these an Inspiration stage may appear. On account on Reality and Inspiration an Experiment stage is operated. The Result and their interpretation follow.

Figure 5.12 show a polytope 4D for RIE methodology.

The direct sequence is:

- S-Reality

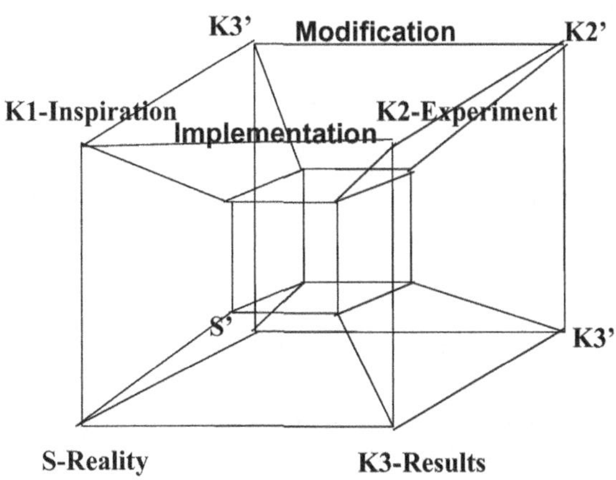

Fig. 5.12 Polytope 4D for RIE

- K1-Inspiration
- K2-Experiment
- K3-Results.

Observe that the four major stages are common to many research or project implementation.

The stage S evaluates some aspects of reality and the objective of projects also. The projects start by asserting a goal, a problem to solve, a product to fabricate, an installation to design and control, a company to build and this is significant for the selection of domain of reality exam and of building elementary bricks.

The stage K1 refers to project ideas, to design of experiment, to available tools.

The stage K2 supposes to realize a prototype, an installation, and to effectively run the research, the production or the project.

The stage K3 refers to data analysis, formal evaluation and field applications.

The reverse sequence is:

- S'-Reality'
- K1'-Inspiration'
- K2'-Experiment'
- K3'-Results'.

Reverse sequence refers to modified, or to post formal stages.

After the direct sequence a conservative strategy may operate the reverse sequence that modifies different stages (Fig. 5.12).

Implementation and modification coexists in duality.

During the dual processes of implementation and modification the differences allowing self-evolvability may be generated.

The direct sequence S → K1 → K2 → K3 corresponds to implementation. The reverse sequence K3' → K2' → K1' → S' corresponds to modification.

In a conservative strategy the modified stages are:

- S'-Reality'
- K1'-Inspiration'
- K2'-Experiment'
- K3'-Results'.

Figure 5.13 shows an example of exploratory or innovative strategy.

Instead of repeating and ameliorating the results of the direct sequence a new sequence is performed.

The reverse sequence is K4 → K5 → K6 → K7.

An example refers to chemical technology. After evaluation of the results the decision may be to change from micro-scale to nano or macro-scale for supporting environment, or from pressure swing absorption PSA to temperature swing absorption TSA and so on.

Direct sequence operates PSA while reverse sequence operates TSA.

The direct and exploratory sequences may be non-optimal but applied in series may offer suitable results.

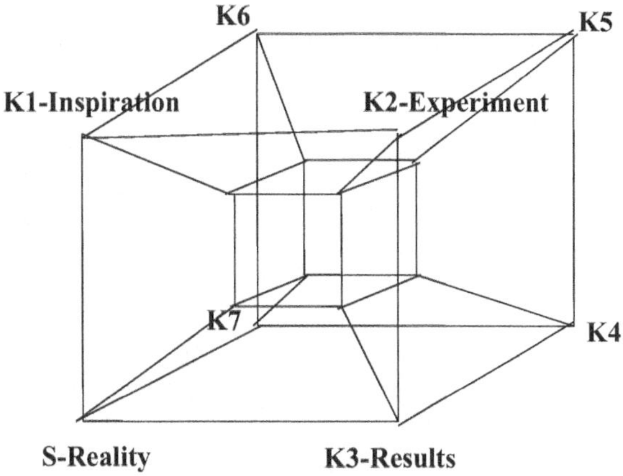

Fig. 5.13 Exploratory polytope 4D for RIE

We may consider a more detailed study of each stage or level.

This means to consider the stages S, K1, K2 and K3 as major stages or tiers in which the sub-stage or sub-levels 0, 1, 2, 3 are identified.

Details or sub-levels may be considered at any of the four levels.

These sub-levels or sub-stages have been introduced in development stage theory by Fischer [7].

They have been denoted 0-singles, 1-mappings, 2-systems, and 3-SoS. Obviously we need to find interpretations of sub-levels for the particular domain considered.

For Reality a natural choice may be: 0-matter, 1-bio, 2-cognitive, 3-intelligent related significance.

The sub-levels 0', 1', 2', 3' denotes modifications or improvements of sub-levels 0, 1, 2 and 3. This means that a conservative development strategy may be selected for sub-levels.

Figure 5.14 shows an exploratory polytope 6D for RIE.

Figure 5.15 identify the reverse sequence stages K4, K5, K6 and K7 according to MHC theory [3].

The new stages of the reverse sequence stage are:

- MS-metasystematic
- P-paradigmatic
- CP-cross-paradigmatic
- MCP-meta-cross-paradigmatic.

For MS stage the scholar can form relationships between systems and understand similarities and differences. An example may be the separation methods.

For P stage the scholar can define larger-picture paradigms. An example may be to complement the physical separations with chemical separations as a new paradigm.

For CP stage the scholar can see relationship between multiple paradigms. An example may be hydro-dynamical, physico-chemical methods.

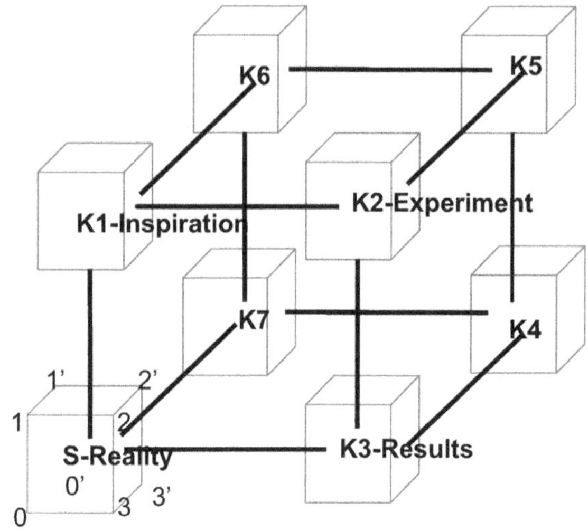

Fig. 5.14 Exploratory polytope 6D for RIE

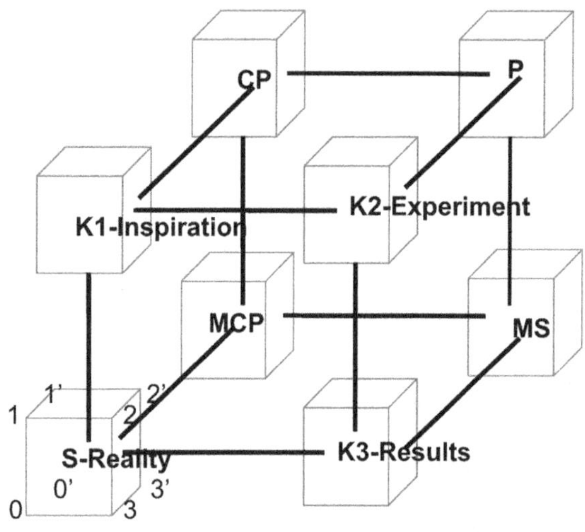

Fig. 5.15 Exploratory post formal polytope 6D for RIE

For MCP stage the scholar should reflect on the similarities and differences between multiple paradigmatic relationships.

References

1. Auzelle, J.P., Morel, G., Panetto, H., Mayer, F.: Using systems of systems engineering to improve the integration of enterprise-control systems. Special issue on Systems Engineering: Best of France. 11/3. Insight J. INCOSE, July, 2008 https://www.incose.org/ProductsPubs/per

iodicals/insight.aspx (2008)
2. Bagajewicz, M.J.: On the integration of process and product design with microeconomics and finances. In: Enpromer 2005, 2nd Mercosur Congress on Chemical Engineering, 4th Mercosur Congress on Process Systems Engineering (2005)
3. Commons, M.L.: Introduction to the model of hierarchical complexity and its relationship to postformal action. World Futures **64**(5–7), 305–320 (2008)
4. Coppens, M.O.: Nature inspired chemical engineering learning from the fractal geometry of nature in sustainable chemical engineering. Fractal Geom. Appl.: A Jubilee Benoit Mandelbrot 2004, **72**, 07–32 (2004)
5. Coppens, M.O.: Scaling-up and-down in a nature-inspired way. Ind. Eng. Chem. Res. **44**(14), 5011–5019 (2005)
6. Coppens, M.O.: A nature-inspired approach to reactor and catalysis engineering. Curr. Opin. Chem. Eng. **1**(3), 281–289 (2012)
7. Fischer, K.W.: A theory of cognitive development: the control and construction of hierarchies of skills. Psychol. Rev. **87**, 477–531 (1980)
8. Garg, M.K., Douglas, P.L., Linders, J.G.: An expert system for identifying separation processes. Can. J. Chem. Eng. **69**, 67–75 (1991)
9. Gerbaud, V., Shcherbakova, N., Da Cunha, S.: A nonequilibrium thermodynamics perspective on nature-inspired chemical engineering processes. Chem. Eng. Res. Des. **154**, 316–330 (2020)
10. Gerbaud, V., Xuereb, C., Coppens, O.: Nature-inspired chemical engineering processes. Chem. Eng. Res. Des. **155**, 200–201 (2020)
11. Hammer, M., Champy, J.: Reengineering the Corporation. Harper Collins, New York (1993)
12. Hartmann, N.: The New Ways of Ontology. Greenwood Press, Westport (1952)
13. Iordache, O., Corriou, J.P., Garrido-Sanchez, L., Fonteix, C., Tondeur, D.: Neural network frames applied for biochemical kinetic diagnosis. Comp. Chem. Eng. **17**(11), 1101–1113 (1993a)
14. Iordache, O., Corriou, J.P., Tondeur, D.: Separation sequencing. Use of information distance. Can. J. Chem. Eng. **71**, 955–966 (1993b)
15. Koutsoukis, N.S., Mitra, G., Lucas, C.: Adapting on-line analytical processing for decision modelling: the interaction of information and decision technologies. Decis. Support Syst. **26**(1), 1–30 (1999)
16. Lepek, D., Coppens, M.O.: Nature-inspired chemical engineering: course development in an emerging research area. In: Proceedings of the ASEE Annual Meeting (2016)
17. Morel, G., Panetto, H., Zaremba, M.B., Mayer, F.: Manufacturing enterprise control and management system engineering: paradigms and open issues. IFAC Annu. Rev. Control **27**(2), 199–209 (2003)
18. Perera, A.S., Coppens, M.O.: Re-designing materials for biomedical applications: from biomimicry to nature-inspired chemical engineering. Philos. Trans. R. Soc. A **377**(2138), 20180268 (2019)
19. Poli, R.: The basic problem of the theory of levels of reality. Axiomathes **12**, 261–283 (2001)
20. Poli, R.: Three obstructions: forms of causation, chronotopoids, and levels of reality. Axiomathes **17**, 1–18 (2007)
21. Robertson, J.L., Subrahmanian, E., Thomas, M.E.,Westerberg, A.W.: Management of the design process: the impact of information modeling. In: Biegler, L.T., Doherty,M.F. (eds.) Fourth International Conference on Foundations of Computer Aided Process Design Conference, AIChE Symposium Series No. 304, 91, pp. 154–165. CACHE Corp. and AIChE (1995)
22. Senge, P.M.: The Fifth Discipline: The Art and Practice of the Learning Organization. Doubleday, New York (1990)

Chapter 6
Industries

6.1 Industry 4.0

Figure 6.1 represents the industry 4.0 as the fourth dimension in a 4D polytope.

Here Ind is the abbreviation for industry.

The direct sequence Ind 0.0 → Ind 1.0 → Ind 2.0 → Ind 3.0 corresponds to main development stages. The reverse sequence Ind 3.0′ → Ind 2.0′ → Ind 1.0′ → Ind 0.0′ corresponds to modified stages.

This means that a conservative strategy for industry development was adapted.

Blurring the lines between different industrial stages, the fourth industrial revolution (industry 4.0) denoted as Ind 4.0 imposes the fusion represented by the center of the 4D polytope.

The 4.0 approach emphasizes cooperation between industry and science to promote closer links between knowledge and skills [19].

The industry 4.0 working groups describe the using of Internet of Things, IoT, technologies, communications, and web services in manufacturing. These tools create networks incorporating the entire manufacturing process that convert factories into smart factories [9, 10, 20]. Linkages include smart machines, and production facilities that feature end-to-end integration, logistic, production, marketing, and service. Industry 4.0 is projected to create closer cooperation between industry partners, suppliers and customers, and between employees, providing new opportunities for mutual benefit [6, 11].

The vision of industry 4.0 is that of significantly higher productivity, efficiency, and self-managing production processes where people, machines, equipment, logistics systems, and work-in-process components communicate and cooperate with each other directly.

A major goal is the application of mass production efficiencies to achieve make-to-order manufacturing of quantity one by leveraging embedded processing and communications. Production and logistics processes are integrated across company boundaries, creating a real-time lean manufacturing ecosystem that is more efficient

O. Iordache, *Polytopic Roadmaps*, Studies in Systems, Decision and Control 368,
https://doi.org/10.1007/978-3-030-75630-7_6

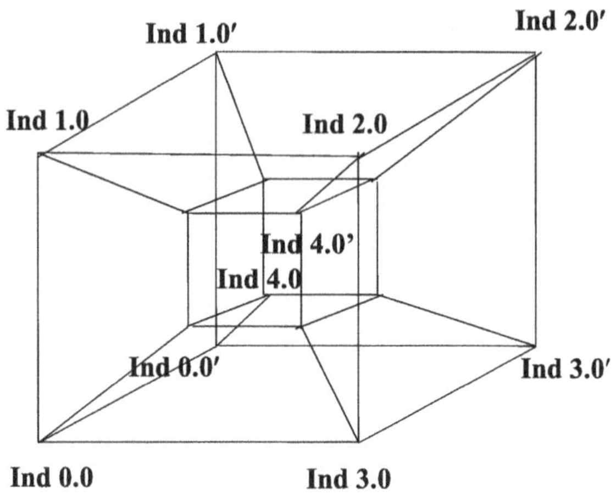

Fig. 6.1 Polytope 4D for industry 4.0

and flexible. This facilitates smart value-creation chains that include all of the life-cycle phases of the product from the initial product idea, development, production, use, and maintenance to recycling.

Networking companies in the supply chain optimize individual production steps and the entire value chain. Comprehensive real-time information enables companies to react during production to the availability of raw materials based on price, quality, and other factors for optimal efficiency. External linkages enable production processes to be controlled across company boundaries to save resources and energy.

The digitization of industry is forecasted to foster new business models and present opportunities for small-and medium-size enterprises. For example, to build low-volume metal parts, companies build virtual 3D models and use direct metal laser sintering, an additive process similar to 3D printing that deposits metal powder layers melted by laser, to create parts. These parts are fully dense metal with specified mechanical properties. Significantly, the 3D and 4D process can produce complex geometries that conventional machining processes are not capable of creating.

Industry 4.0 systems capture a wide range of data that can be used to improve performance and productivity with the application of analytics. Analytics are used in a number of ways, including real-time predictive maintenance, which helps manufacturing companies avoid interruption of production by unplanned machine failures on the factory floor-directly improving asset utilization. An application of analytics is optimization of production operations, improving productivity and energy efficiency.

Industry 4.0 concept was introduced in 2011 and has been discussed since. It is being interpreted as the fourth industrial revolution which is changing the way we work, we learn, we think about production or industrial manufacturing, the skills and competencies that are demanded on the industrial job market. The understanding of industry 4.0 can vary based on focused sectors, companies industries or countries.

Different countries have a vision and a stratagem concerning the fourth industrial revolution. Examples of such visions include Industrie 4.0 in Germany, Smart Manufacturing Leadership Coalition and Advanced Manufacturing Partnership in USA,

Table 6.1 Digital manufacturing initiatives

Country	Year	Name
Australia	2013	Next wave of manufacturing
Belgium	2013	Made different
Canada	2015	Smart manufacturing, Industrie 4.0
China	2015	Made in China 2025
Denmark	2012	Made
France	2015	Industrie du Futur
Germany	2011	Industrie 4.0
India	2014	Make in India
Italy	2016	Fabbrica intelligente, Industria 4.0
Japan	2015	Industrial Value Chain Initiative
Netherlands	2014	Smart Industry
South Korea	2015	Korea Industry 4.0 Association
Spain	2015	Industria Conectada 4.0
Sweden	2014	Produktion 2030
UK	2014	High Value Manufacturing
USA	2012	Advanced Manufacturing Partnership

Made in China 2025 in China, Smart Industry in Netherlands, Fabbrica Intelligente and Industria 4.0 in Italy and so on.

Table 6.1 shows different manufacturing initiatives and the year of public presentation.

Note that the Platform Industrie 4.0 was established in Germany in 2012 while the global Industrial Internet Consortium, IIC, was established in USA in 2014.

6.2 Industry 5.0

There are different visions for industry 5.0. Some users argue that while industry 4.0 is essentially about connecting devices together, industry 5.0 is about collaboration between humans and machines on the factory floors [3, 8].

According to Østergaard [12], industry 5.0 is the return of the human touch on the factory floors. Some users claim that while Europe initiated the fourth industrial revolution, North America and East Asia are distinctively positioned to deal with the following industrial revolution stages, 5.0 and beyond.

Figure 6.2 illustrates the use of the polytope 5D for industry 5.0.

It is considered that the human actions and the industry have similar 4D frames.

The thick line connecting the two polytopes 4D corresponding to industry and human actions ensures that the resulting framework is a 5D polytope. The dimension is $1 + 4=5$.

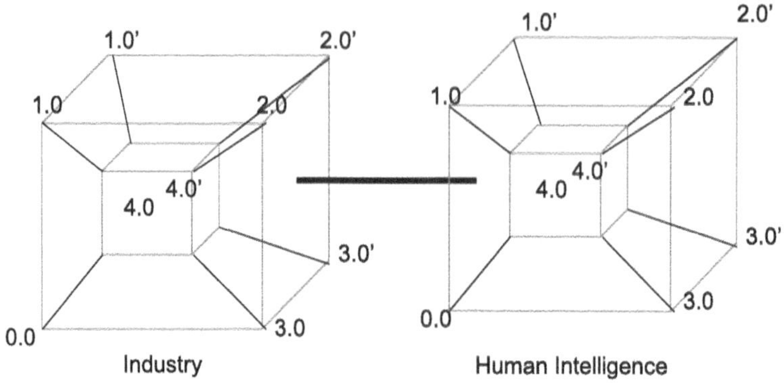

Fig. 6.2 Polytope 5D for industry 5.0

Here 1 corresponds to the line and 4 to the hypercubes 4D.

Two industrial revolutions, 4.0 and 5.0, so close to each other have been treated as one by some users. Therefore, Industry 4.0 may incorporate both connectivity and man-machine collaboration [8].

In practice, artificial intelligence, AI, tools complement the human intelligence.

One of the pillars of Industry 4.0 is robotics and AI. A natural extension of robotics and AI is man-machine collaboration. Therefore just the introduction of human-machine collaboration on the factory floors is not enough for supporting the case for a new industrial revolution for some users.

Industry 5.0 is related to bionics, biomimicry or biomimetics and nature inspired engineering [2, 7, 13].

Bionics is the imitation or abstraction of the inventions of nature [16].

The human intelligence endorses the inspiration and this may increase the dimensionality from 4D to 5D that is from 4.0 to 5.0 as shown in Fig. 6.2.

Reality Inspired Engineering, RIE is a basic method to increase dimensionality. From this perspective industry 5.0 appears as an iteration of industry 4.0 and the iteration process may be continued on the way to polytopic industry.

Bio-economy is the production of renewable biological resources and the conversion of these resources and waste streams into value added products, such as food, feed, and biology based products and bio-energy. It includes agriculture, forestry, food and pulp and paper production, as well as parts of chemical, biochemical, biotechnological and energy industries. Its sectors have a strong innovation potential due to their use of a wide range of sciences (life sciences, agronomy, ecology, food science and social sciences), enabling and industrial technologies (biotechnology, nanotechnology, information and communication technologies, and engineering) .

Bio-economy is essential in achieving a self-sustainable economy [18].

Smart use of biological and bio-inspired resources for industrial purposes will help achieving a balance between industry, economy and ecology.

In the research agenda, several fields of action are given priority. These are:

- Securing global nutrition
- Ensuring sustainable agricultural production
- Producing healthy and safe foods
- Using renewable resources for industry
- Developing biomass-based energy carriers
- Developing bio-inspired engineering

Biology trend, the guiding principle of the bio-economy, has the potential to create a fundamental change in industry. Therefore, bio-economy will be a central part of next industrial revolution.

An ecological vision for industry 5.0 is set forth by Rada [14, 15]. It is considered that the priority of industry 5.0 is to utilize efficiently workforce of machines and people, in synergy with the environment.

The theme of this vision is industrial up-cycling.

This vision focuses on waste prevention. Furthermore, it was pointed out that we need to turn to human element in the manufacturing process. The current digitization trend that is the effort to embed 1 s and 0 s into any living organism is challenged [14].

Industry 5.0 includes 6R methodology. The 6R are:

- Recognize: We need to recognize the opportunities offered by industrial up-cycling. An awareness is the first required step.
- Reconsider: We need to evaluate and reconsider our business and manufacturing processes. A redesign of processes to realize the benefits of industrial up-cycling is an essential step.
- Realize: After recognition of the opportunities and reconsideration of business processes, we need to realize the business process improvement or innovation.
- Reduce: Reducing the use of resources to achieve efficient outcomes is the essence of the methodology.
- Reuse: Reusing the materials considered as useable prior to process improvement is also at the center of the methodology.
- Recycle: Recycling as much as possible is one of main expected outcomes of the industrial up-cycling.

The ideal situation for industry is the zero waste.

The 6R methodology actually defines a business improvement model. Depending on the specific case, it can be considered as a business process modification or an exploratory innovation. Therefore, the 6R methodology is subject to the rules, assumptions, and dynamics of process improvement efforts, that is to reverse sequence in polytopic approach. The methodology may support an exploratory strategy.

In only a few years after introduction of industry 4.0, various visionaries started discussing industry 5.0. These visionaries also pointed out some unexplored potentialities for industry 4.0 and proposed industry 5.0 to overcome the possible constraints of industry 4.0. Previous industrial revolutions occurred naturally unlike industry 4.0, which is formally defined and forced upon the industry. It is possible to argue that this artificial revolution start is precipitate. To call a concept an industrial revolution, we need to observe a widespread change both in industries, businesses, and society. Currently, with its current definition, industry 4.0 should actually be considered just as a far-reaching proposal. It is evident that industry 4.0 is still under development including at the level of conceptualization.

Smart mass production is one of the goals of industry 4.0. Sustainability is main theme in industry 5.0 proposals. Note that sustainability and mass production are not mutually exclusive. Therefore, combining these two goals or themes and redefining the next industrial revolution may be a better approach. As a result, the objective of the next industrial should be the self-evolvable and self-sustainable smart production.

6.3 Industry 4.0 Toolbox

Industry 4.0 is increasingly entering the industrial landscape of Small and Medium Enterprises (SME). However, the right introduction to the subject for the affected enterprises proves to be difficult. For this reason, a generic procedure to introduce industry 4.0 in SME was developed. It is a roadmap towards industry 4.0 and beyond 4.0.

The purpose of this model is to give SMEs guidance on how to address industry 4.0 in order to identify enterprise-specific technology solutions, to optimize existing processes, to advance existing business models and to exploit new business models [1, 19].

The development of toolboxes for engineering follows the Guideline Industry 4.0 set by the VDMA (*Verband Deutscher Maschinen- und Anlagenbau*).

To meet the guidelines, applications are firstly defined and divided in development stages. Relevant applications are identified based on the defined engineering tasks and potentials areas. Furthermore, the applications are classified respecting the categories of engineering areas. The most important applications have to be selected and used in the toolbox.

Toolbox tables are structured in applications vertically (y-axis) and development stages horizontally (x-axis).

Application displays industry 4.0 themes. Every application is broken down in five technological and sequential stages. The lowest level 0.0 correspond to the conventional usually non-digitalized stage. The highest level 4.0 represents the industry 4.0 perspective.

Several applications may be grouped in areas or groups. Examples of areas of applications are Product, Production, Business Models, Intra-logistics, Assembly and Security.

Table 6.2 shows examples of industry 4.0 toolboxes.

Application areas describe topics of interest in the context of industry 4.0, for industrial product development as well as for industrial production. The first four rows in Table 6.2 refer to product, while the last four rows refer to production.

Table 6.2 Industry 4.0 toolbox; Product and Production

Application	0.0	1.0	2.0	3.0	4.0
Integrationsensors/actuators	No sensors/actuators	Sensors/actuators integrated	Sensors reading processed by system	Data evaluated analyzed by system	System independently respond based on data analysis
Communication connectivity	No interface	I/O signals	Field bus interface	Ethernet interface	Product access to internet
Monitoring	No monitoring	Failure detection	Recording operating conditions	Prognosis of condition	Independent control Measures
Product related IT service	No services	On line portals	Service direct via product	Independent performed service	Complete integration in infrastructure of IT services
Data processing	No processing	Storage data	Analyze data	Evaluation for planning control	Automatic process planning control
Company-wide networking	No networking	Info via mail phone	Rule for data exchange	Interdivision inked data servers	Interdivision fully networked IT solutions
ICT infrastructure	Info via mail phone	Central data servers	Internet portals	Automated Info exchange	Suppliers/customers fully integrated
Efficiency with small batches	Rigid production systems	Flexible production identical parts	Flexible production modular design	Component driven flexible production	Component driven modular production value-adding networks

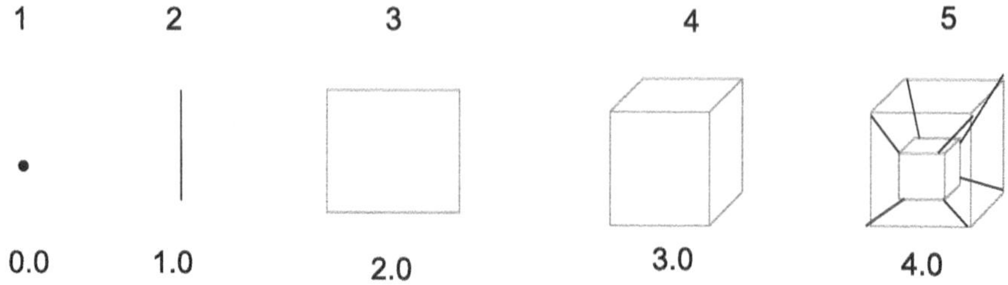

Fig. 6.3 Polytopic Roadmaps: stages and illustrations

Performance levels describe technological approaches which are structured into a sequence reaching from none or low level industry performance labeled 0.0 to high level industry 4.0 performance.

The development stages may be designated from 1 to 5 or from 0.0 to 4.0.

Figure 6.3 shows these stages and proposes common illustrations for different applications. Stages are illustrated by dots, lines, squares, cubes and 4D hypercubes.

We are faced with Polytopic Roadmaps for development stages.

It is also a first row in a Periodic Table that should be completed for different application areas. Figure 6.3 shows five stages since an extra stage 4.0 was included.

This can be considered as an extension of the development skill theory [4] where only four stages have been considered.

Fischer illustrates the sub-levels of development by dots, lines, squares and cubes.

The sub-levels names are singles, mappings, systems and SoS. The supplementary sub-level shown in Fig. 6.3 illustrates system of systems of systems, SoSoS.

The first column from Table 6.2 corresponds to the specific applications.

Consider for instance that the application is the integration of sensors and actuators. The development stage starts with the product having no use of sensors and actuators (stage 0.0). Industry 4.0 vision is that the product responds independently based on gained data (stage 4.0) . The intermediary stage (1.0)-sensors integrated stage (2.0)-sensor reading and stage (3.0)-data evaluated for this application are established on the basis of the expertise in the domain. Practical expertise allows defining the qualitative progress steps for the specific domain of application. These steps should be associated to singles, mappings, systems, SoS and SoSoS.

The next application in Table 6.2 is the communication and connectivity. The development starts with the product having no interfaces for communication (stage 0.0). Industry 4.0 vision is that of the product having access to the internet (stage 4.0) .

Another example of application is the monitoring. The development starts with no monitoring for the product (stage 0.0) . Industry 4.0 vision is the independently adopted control measures of the product (stage 4.0) .

The last example of application is the product-related IT services. The development stage starts with no services. Industry 4.0 vision is that of the complete integration into an infrastructure of IT services.

Figure 6.4 shows the polytope 4D for industry 4.0 toolbox.

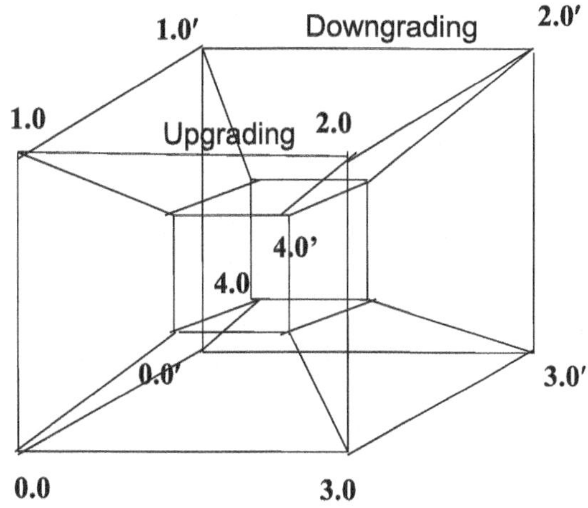

Fig. 6.4 Polytope 4D for industry 4.0 toolbox

It is valid for any specific application.

The 4D polytope outlined two sequences for evolution the direct sequence S → K1 → K2 → K3 that corresponds to the upgrading stages and the reverse sequence K3′ → K2′ → K1′ → S′ that corresponds to the downgrading modified stages.

"Self" is coupling the two sequences.

This is a conventional or conservative strategy of development. The direct acquired development is questioned and modified by the reverse sequence.

The stages 0.0, 1.0, 2.0, 3.0, 4.0 may be associated to S, K1, K2, K3 and "Self" in 4D polytope. They correspond to upgrading.

The stages 0.0′, 1.0′, 2.0′, 3.0′, 4.0′ correspond to downgrading and may be associated to modified stages that is to S′, K1′, K2′, K3′ and "Self′". The modifications means that once the application arrived at 4.0 all the preliminary stages may be modified (digitalized in many situations). For instance, if we refer to monitoring, a stage like failure detection 1.0, should benefits from digitalizing and be modified to 1.0′.

The high dimensional polytope for Industry toolbox may be defined.

We can select the four applications referring to the product area.

The application and the abbreviations are:

- Integration-Int
- Communication-Com
- Monitoring-Mon
- IT Service-IT

They correspond to the first four rows in the Table 6.2.

Figure 6.5 show the Polytopic Roadmaps for industry 4.0 toolbox-product

The Polytopic Roadmaps from Fig. 6.5 show a roadmap to Industry 6.0 since the highest dimension polytope shown in Figs. 6.6 and 6.5 is a 6D polytope, a square of hypercubes. The dimension is 2 + 4=6. Here 2 corresponds to the square and 4 to the hypercubes 4D connected by the square.

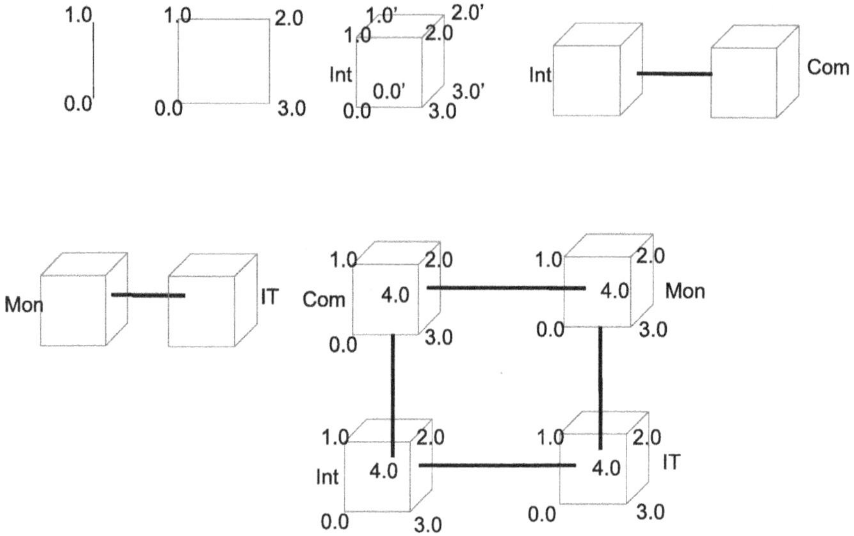

Fig. 6.5 Polytopic Roadmaps for industry 4.0 toolbox-product

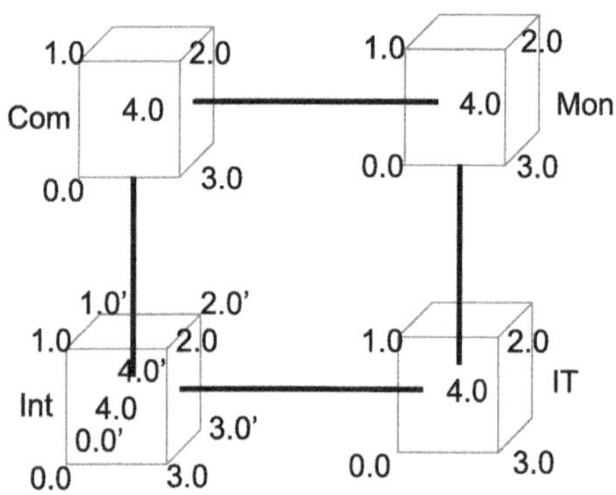

Fig. 6.6 Polytope 6D for industry 4.0 toolbox-product

The roadmap starts from 0.0. The first step in development is 1.0. To this step the 1D framework is associated. For that application the development continues to include 2.0, 3.0 and 4.0 and more than this the modified stages 0.0′, 1.0′, 2.0′, 3.0′, and 4.0′. Some stages have not been included in Fig. 6.5 to avoid diagram complexity. With these we complete a first level or stage of development Integration-Int. The following levels Com. Mon, and IT and couplings for instance as Int-Com or Mon-IT will complete the development towards industry 6.0 for product.

Figure 6.6 shows more details of the polytope 6D for industry 4.0 toolbox-product.

To any application domain we associate the stages 0.0 to 4.0 and the modified stages 0.0′ to 4.0′.

Figure 6.7 shows the polytope 7D for industry 4.0 toolbox-product.

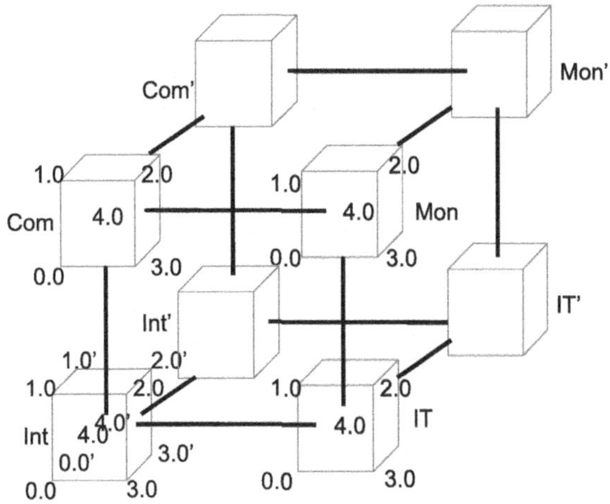

Fig. 6.7 Polytope 7D for industry 4.0 toolbox-product

Using the thick hyper-lines, the 7D cube is shown as cube of hypercubes. The dimension is 3 + 4=7. Here 3 correspond to the cube represented by thick lines and 4 to the hypercubes, 4D connected by that cube.

The developed 7D polytope represents steps towards Industry 7.0 since the associated polytope is 7D.

To arrive at this stage of industry evolution, the basic levels, that is the applications should be modified,

We adopted a conservative strategy for main levels development.

The modified basic levels are denoted as follows:

- Integration'-Int'
- Communication'-Com'
- Monitoring'-Mon'
- IT Service'-IT'

Observe that there are two scales for levels modification.

The large scale modifies Int to Int' while at a smaller scale, inside Int we can modify 1.0 to 1.0′.

After the product study, we will select from Table 6.2 four applications domains referring to production area. The applications and the abbreviations are:

- Data Processing-DP
- Networking-Net
- ICT infrastructure-ICT
- Efficiency-Eff

They are the next group of four rows in the Table 6.2.

Figure 6.8 shows the Polytopic Roadmaps for industry 4.0 toolbox-production

Figure 6.8 shows as final framework a square of hypercubes. This is a polytope 6D associated to industry 6.0.

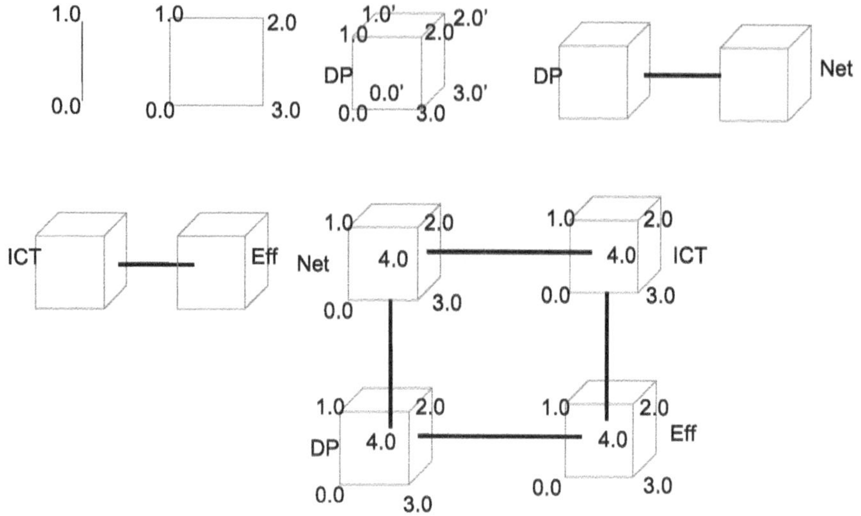

Fig. 6.8 Polytopic Roadmaps for industry 4.0 toolbox-production

Figure 6.9 shows the polytope 6D for industry 4.0 toolbox-production.
Figure 6.10 shows the polytope 7D for industry 4.0 toolbox-production
Using the thick hyper-lines, the 7D cube is shown as cube of hypercubes.
The developed 7D polytope represents steps towards Industry 7.0.
We adopted a conservative strategy for large stage development.
The modified basic levels are:

- Data Processing'-DP'
- Networking'-Net'
- ICT Infrastructure'-ICT'
- Efficiency-Eff'

Fig. 6.9 Polytope 6D for industry 4.0 toolbox-production

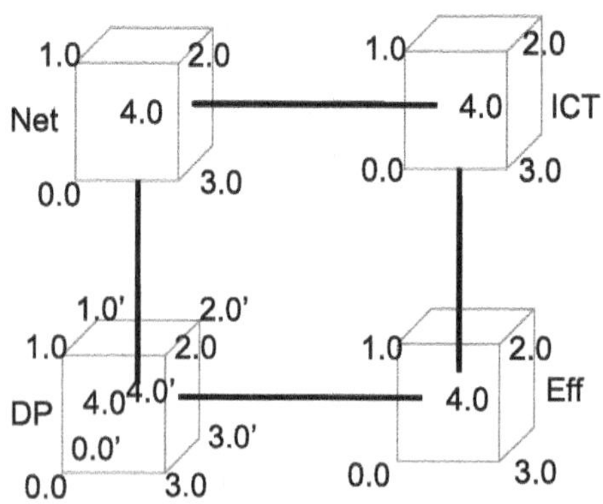

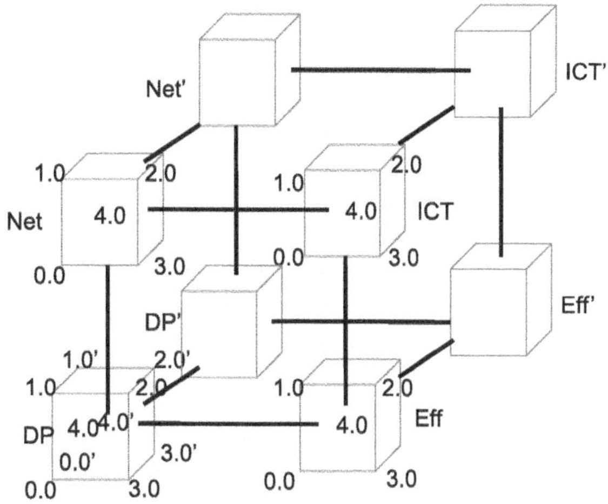

Fig. 6.10 Polytope 7D for industry 4.0 toolbox-production

The maturity index calculated for industry 4.0 toolbox may be completed by the calculus of dimension [17].

Table 6.3 contains industry 4.0 toolbox for Business Model.

The first column from Table 6.3 corresponds to the specific applications.

Consider for instance the application of the Connection Value-added Process denoted CVP. The development stage starts with no connections (stage 0.0). Industry 4.0 vision is the open for all market participants (stage 4.0) .

Another application in Table 6.3 is the Consistent Digital Engineering denoted CDE.

The development stage starts with the product having no connection (stage 0.0) . Industry 4.0 vision is the smart factory (stage 4.0) .

Another example of application is the Product Development denoted PD.

The development starts with non digital (stage 0.0) . Industry 4.0 vision is the open source concept (stage 4.0) .

The next example of application is the Demand Availability of Machines denoted DAM. The development stage starts with no data connection (stage 0.0) . Industry 4.0 vision is that of querying free capacity of external machines (stage 4.0) .

The stages 0.0, 1.0, 2.0, 3.0, 4.0 may be associated to S, K1, K2, K3 and "Self" in 4D polytope. They correspond to upgrading.

The stages 0.0′, 1.0′, 2.0′, 3.0′, 4.0′ correspond to downgrading and may be associated to modified stages that is to S', K1′, K2′, K3′ and "Self'".

We adapted a conservative strategy for large stage development.

The high dimensional polytope for Industry toolbox-business model may be defined.

We take into consideration the four applications referring to the business model area:

- Connection Value-added Process-CVP
- Consistent Digital Engineering-CDE

Table 6.3 Industry 4.0 toolbox: business model

Application	0.0	1.0	2.0	3.0	4.0
Connection value-added process	No connections	Closed model	Connecting individuals involved	Connecting all participants	Open for all market participants
Consistent digital engineering	Non digital process development	Digitization and standardization	Data management	Digital factory	Smart factory
Product development	Non digital by local teams	Digital processes	Outsourcing development activities	Synchronous collaboration	Open source concept
Demand availability of machines	No data connection	M2M condition monitoring	Prognosis of functional condition	Automatic execution of service	Querying free capacities of external machines

- Product Development-PD
- Demand Availability of Machines-DAM

They are recorded as the first column in the Table 6.3.

Figure 6.11 show the Polytopic Roadmaps for industry 4.0 toolbox-business model.

The basic levels in Fig. 6.11 are denoted as follows: CVP, CDE, PD and DAM.

The Polytopic Roadmaps shows a roadmap to Industry 6.0 since the highest dimension polytope shown in Fig. 6.11 is a 6D polytope, a square of hypercubes.

Figure 6.12 shows the polytope 6D for industry 4.0 toolbox-product.

The basic levels in Fig. 6.12 are CVP, CDE, PD and DAM.

To any application domain we associate the stages 0.0 to 4.0 and the modified stages 0.0′ to 4.0′.

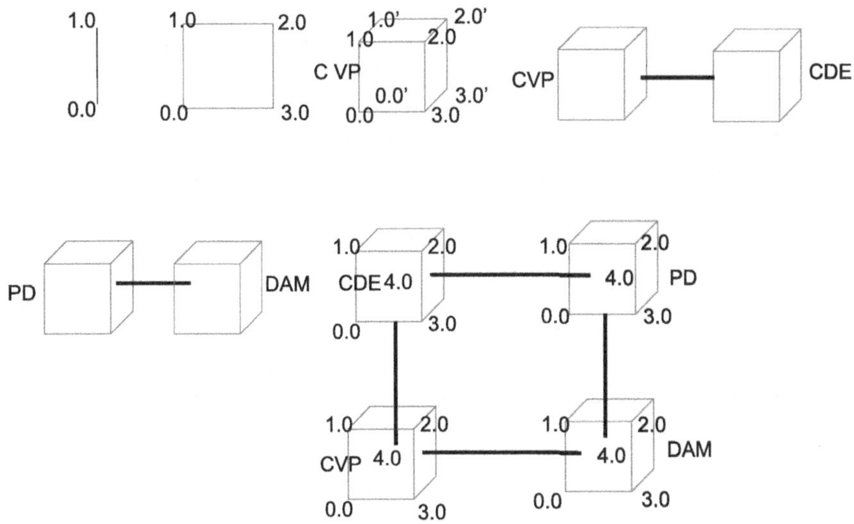

Fig. 6.11 Polytopic Roadmaps for industry 4.0 toolbox-business model

Fig. 6.12 Polytope 6D for industry 4.0 toolbox-business model

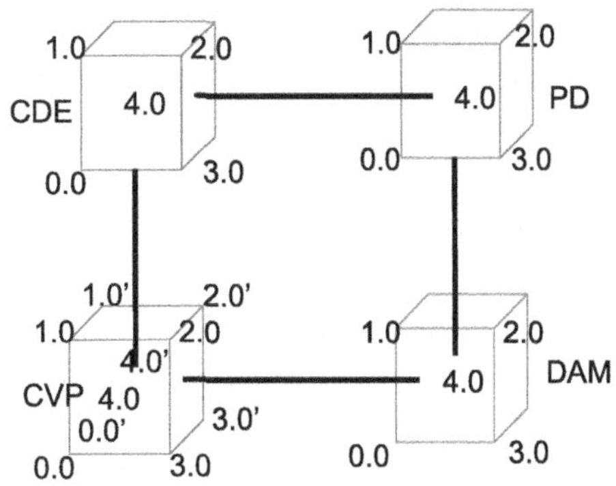

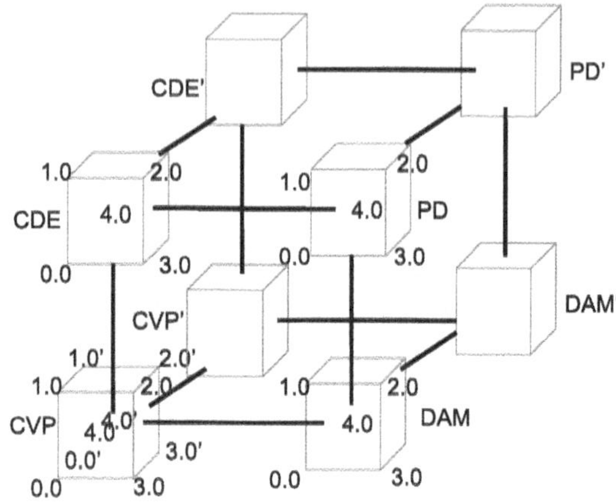

Fig. 6.13 Polytope 7D for industry 4.0 toolbox-business model

Figure 6.13 shows the polytope 7D for industry 4.0 toolbox-product.

Using the thick hyper-lines, the 7D cube is shown as cube of hypercubes. The dimension is 3 + 4=7.

The developed 7D polytope represents steps towards industry 7.0 since the associated polytope is 7D. To arrive here, the basic levels, that is the applications should be modified.

We adopted a conservative strategy for large stage development.

The modified basic levels are denoted: CVP', CDE', PD' and DAM'.

Obviously it is possible to adopt an exploratory strategy and surpass the modification at basic level or large scale.

This modification may be replaced with a new set of applications.

For instance, we can put together the product and the production applications domains shown in Table 6.2.

Figure 6.14 shows the polytope 7D for industry 4.0 toolbox-product-production.

The applications Int, Com, Mon and IT pertains to the product while the applications DP, Net, ICT and Eff pertains to production.

This corresponds to an innovative or exploratory strategy of industrial development.

The reverse stages are radically new since they do not refer to the same application domains. Instead of a conservative strategy of product modifications an exploratory strategy going from product to production is adopted.

6.4 Polytopic Industry

Figure 6.15 shows a Periodic Table for polytopic industry.

The Periodic Table contains columns (groups) and rows (periods).

Fig. 6.14 Polytope 7D for industry 4.0 toolbox-product-production

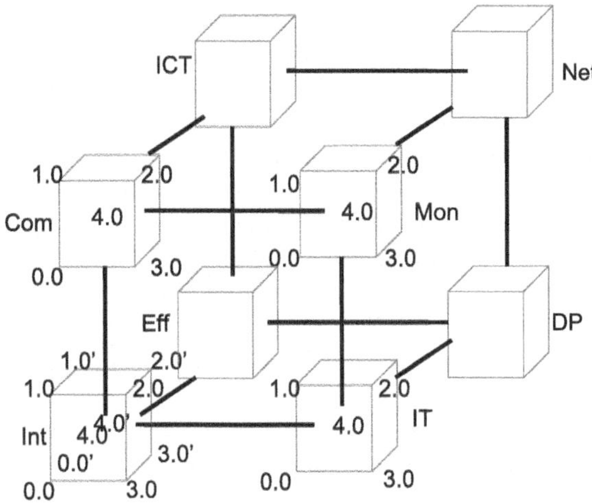

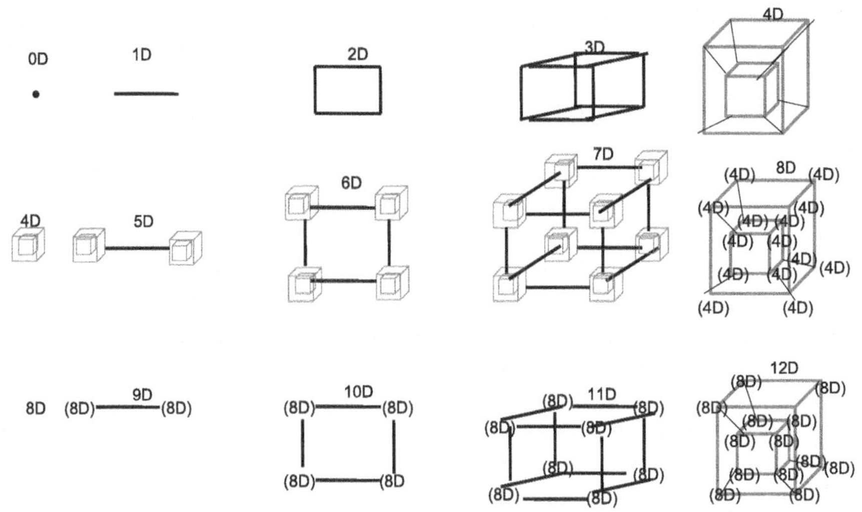

Fig. 6.15 Periodic Table for polytopic industry

The columns correspond in succession to dots, lines, squares, cubes and 4D-hypercubes that is to 0D, 1D, 2D, 3D and 4D frames. The columns highlight the dot, the line, the square and the cube and the hypercube as patterns or skeletons.

The rows use as basic bricks the 0D (dots) in the first row, 4D-cubes in the second row and 8D (hypercube of hypercubes) in the third row.

Any row show the same period: dot, line, square, cube, hypercube.

The next row of the Periodic Table unincluded in this figure will show 12D, 13D, 14D, 15D and 16D frames.

The nD frame is associated to industry n.0, for instance the 4D frame is associated to industry 4.0, the 5D frame to the industry 5.0 and so on. The 6D and 7D frames based on Industry 4.0 toolbox are illustrated by the Periodic Table.

The labels industry 6.0, industry 7.0 and so on may be naturally associated to such high dimensional reference architectures.

The toolbox allows increasing the industry level according to the number of applications selected for evolution.

For the same enterprise different levels of industry may coexist.

There exist strategies towards polytopic industry independent of the industry 4.0 toolbox.

The strategy based on organization is an example that will be presented in the following.

Digital convergence and divergence is less a process of simple addition and subtraction as it is a complete system reconfiguration. Organizations will need to master the act of growing the core of their companies while scaling for new business models.

The idea encompasses how everything from consumer transactions to production chains is undergoing fundamental changes-and how human skills play an integral role.

From the wheel to the web, technology has always made work easier, freeing people to focus on more interesting challenges. In spite of rising automation, none of these combinations will see their full potential if companies don't also invest in the designers, researchers, and entrepreneurs that deliver innovation.

Using polytopic approaches means that several extra dimensions can be added to a project. Where traditional construction relies on 2D drawings and printed reports for information sharing, 3D digital models of a building or plant allow project teams to view a design as if it has been built already. The aspect of time makes it 4D.

The construction process can then be viewed as a digitized video, which is very useful in planning design, construction, procurement and logistics and for detecting and preventing clashes between sub-schedules or design elements.

Adding the cost aspect could makes it 5D, where financial data about the elements of design, construction and the consequences of a design change can be visible in real time. Sustainability goals make it 6D. Strategic decisions regarding maintenance can be made by adding maintenance data, making the project 7D.

We may consider a more detailed study of each stage or level.

This means to reevaluate the levels as major levels or tiers in which the sub-levels 0, 1, 2, 3, 4 are identified.

These sub-levels or sub-stages have been introduced in development stage theory by Fischer [4]. The 5th sub-level denoted here 4 may be introduced.

The sub-levels have been denoted 0-singles, 1-mappings, 2-systems, 3-SoS. 4-SoSoS.

Obviously we need to find interpretations of sub-levels for the particular domain considered.

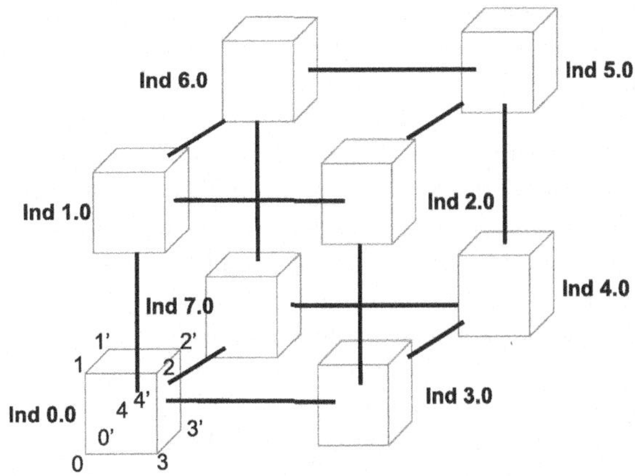

Fig. 6.16 Exploratory polytope 7D

The sub-levels 0′, 1′, 2′, 3′, 4′ denotes modifications or improvements of sub-levels 0, 1, 2, 3 and 4. This means that a conservative development strategy may be selected for sub-levels.

Figure 6.16 represents the industry 7.0 in an exploratory 7D polytope.

Here Ind is the abbreviation for industry.

The direct sequence Ind 0.0 → Ind 1.0 → Ind 2.0 → Ind 3.0 corresponds to main development stages. The reverse sequence Ind 4.0 → Ind 5.0 → Ind 6.0 → Ind 7.0 corresponds to exploration stages.

This means that we adapted an exploratory strategy for industry development levels.

Digital platforms (DP) and digital ecosystems offer another strategy for polytopic industry implementation [5].

The polytopic approach reduces failure costs, eliminates design changes and shortens construction time. The polytopic capability offers significant advantages when investing in new facilities. Modern industrial processes can be complex, partly due to the number of engineering disciplines required.

The polytopic approach enables all the information to be shared during the design process and enhances communications across the project team wherever they are in the world.

At the level of industry 4.0 coordinated operation of digital networking enterprise is achieved via use of a software and communication infrastructure called also digital platform, DP. The main objective of this platform is to constitute, for the network components, common information and communication space as well as runtime environment. Such view of DP looks too limited for the next generation of enterprises comprising of autonomous subsystems, which should be capable to proactive behavior, adaptive planning and on-line de-planning, communicating the enterprise units to employees.

Creating DP with extended capabilities is the key issue of the next step of digital economy formulated as polytopic industry. More particularly, the capabilities of digital enterprise DP are to unite enterprise subsystems into a common semantic information space and provide for them access to the rich set of reusable applied services intended to support for distributed resource planning and control in real time.

DP of polytopic industry should also provide the networked enterprise units with the standard access to the cloud resources and services and to data perceived by external smart sensor networks. A digit role of transformer ecosystem is a distributed, adaptive, open socio-technical system with properties of self-organization, scalability and sustainability inspired from natural ecosystems. Digital ecosystem models are informed by knowledge of natural ecosystems, especially for aspects related to competition and collaboration among diverse entities. It is worth to note that exactly the same necessary features were claimed for an enterprise of polytopic industry can be viewed as a particular case of digital ecosystem, and enterprise DP—as a particular case of ecosystem DP. To be more precise, the enterprise DP should play the role transformer of a digital enterprise to the digital ecosystem of smart services. The ecosystem of smart services should hold the following basic new features [5]:

- Openness—ability to introduce new services into ecosystem, with no stop or restart of it
- Distributed framework—ability of all services to operate autonomously, continuously, in parallel and asynchronously

The service calls can be initiated by either users, or by other services, or the calls can be initiated proactively by internal events generated depending on the system state, criteria values, and decision-making policy.

- Adaptability-the ecosystem ability to change its structure or functions in dynamic fashion in order to increase own efficiency or, in a narrower sense, the ability of ecosystem to adaptively respond to events via change of previous decision or adaptively revisit the existing plan due to change of situation
- Self-organization-the ability of ecosystem of services to pro-actively create local connections and review them when the situation changes
- Service competition and cooperation is the capability of the ecosystem service requester to choose among the services provided by various service suppliers. In contrast to traditional closed hierarchical centralized and sequential systems for enterprise management, the considered enterprise ecosystem is built from autonomous intelligent component capable for proactive situation analysis and resource management, to interact on joint actions, to respond to incoming real-time events through on-line revisiting the current plans to achieve ultimate goals. Implementing of enterprise ecosystem with such features requires development of a fundamentally new DP. As a rule, ecosystem of smart services is composed of a large or even huge number.

References

1. Anderl, R., Picard, A., Wang, Y., Fleischer, J., Dosch, S., Klee, B., Bauer, J.: Guideline industry 4.0—guiding principles for the implementation of industry 4.0 in small and medium sized businesses. In: VDMA Forum Industry 4.0, Frankfurt (2015)
2. Coppens, M.O.: A nature-inspired approach to reactor and catalysis engineering. Curr. Opin. Chem. Eng. **1**(3), 281–289 (2012)
3. Demir, K.A., Cicibas, H.: Industry 5.0 and a critique of industry 4.0. In: Proceedings of the 4th International Management Information Systems Conference, Istanbul, Turkey, 17–20 October (2017)
4. Fischer, K.W.: A theory of cognitive development: the control and construction of hierarchies of skills. Psychol. Rev. **87**, 477–531 (1980)
5. Gorodetsky, V., Larukchin, V., Skobelev, P.: Conceptual model of digital platform for enterprises of industry 5.0. In: International Symposium on Intelligent and Distributed Computing. Springer, Cham, pp. 35–40 (2019)
6. Hwang J.S.: The Fourth industrial revolution (industry 4.0): intelligent manufacturing. Surf. Mount Technol. (SMT) Mag. 10–15 (2016)
7. Iouguina, A., Dawson, J.W., Hallgrimsson, B., Smart, G.: Biologically informed disciplines: a comparative analysis of bionics, biomimetics, biomimicry, and bio-inspiration among others. Int. J. Des. Nature Ecodyn. **9**(3), 197–205 (2014)
8. Johansson, H.: Profinet Industrial Internet of Things Gateway for the Smart Factory. Master's Thesis in Embedded Electronic System Design, Department of Computer Science and Engineering, Chalmers University Of Technology, University Of Gothenburg Gothenburg, Sweden (2017)
9. Kagermann, H., Lukas, W.-D., Wahlster, W.: Industrie 4.0: Mit dem Internet der Dinge auf dem Weg zur 4. industriellen Revolution, VDI Nachrichten, **13** (2011)
10. Kagermann, H., Wahlster, W., Helbig, J. (eds.): Recommendations for implementing the strategic initiative Industrie 4.0: Final report of the Industrie 4.0 Working Group (2013)
11. Lee, J., Bagheri, B., Kao, H.A.: A cyber-physical systems architecture for industry 4.0-based manufacturing systems. Manuf. Lett. **3**, 18–23 (2015)
12. Østergaard, E. H.: Industry 5.0—return of the human touch. Retrieved from https://blog.universal-robots.com/industry-50-return-of-the-human-touch (2016)
13. Perera, A.S., Coppens, M.O.: Re-designing materials for biomedical applications: from biomimicry to nature-inspired chemical engineering. Philos. Trans. R. Soc. A **377**(2138), 20180268 (2019)
14. Rada, M.: Industry 5.0—from virtual to physical, 1 December (2015). https://www.linkedin.com/pulse/industry-50-from-virtual-physical-michael-rada (2015)
15. Rada, M.: Industry 5.0 definition, 3 Feb 2017. Retrieved from https://www.linkedin.com/pulse/industrial-upcycling-definition-michael-rada (2017)
16. Rendall, M.: The New Terminology: CRO and Industry 5.0. Retrieved from https://www.automation.com/automation-news/article/the-new-terminology-cro-and-industry-50 (2017)
17. Sachsenmeier, P.: Industry 5.0—the relevance and implications of bionics and synthetic biology. Engineering **2**(2), 225–229 (2016)
18. Schuh, G., Anderl, R., Gausemeier, J., ten Hompel, M., Wahlster, W.: Industrie 4.0 maturity index. Managing the digital transformation of companies. Munich: Herbert Utz. (2017)
19. Schütte, G.: What kind of innovation policy does the bioeconomy need? New Biotechnology 27 Apr 2017, 40(Pt A): 82–86 (2017)
20. Schwab, K.: The Fourth Industrial Revolution. World Economic Forum, Geneva (2016)
21. Zuehlke, D.: Smartfactory–from vision to reality in factory technologies. IFAC Proc. **41**(2), 14101–14108 (2008)

Chapter 7
Societies

7.1 Polytopic Society

Besides manufacturing processes or factory automation, industrial revolutions have an impact in domains such as education, healthcare, agriculture, economy and so on.

So far, a remarkable initiative is society 5.0 in Japan [13]. Society 5.0 uses the technology as basis towards a human machine symbiosis throughout the society. It targets to improve the life of every person.

Super smart society or society 5.0, is a concept introduced by the Japanese government under which various twenty-first century challenges are addressed: climate change, food security, the limited availability of natural resources and sustainable development, the ageing population.

Instead of the technology-driven approaches pursued thus far, Japan's present strategy incorporates a human-centered approach. The goal is an inclusive society in which everyone shares the same opportunities and that guarantees work, economic growth and well-being for all.

The fundamental technologies of the concept are artificial intelligence, robotics and the Internet of Things. The construction and use of databases are to be designed in order to provide people with important new services. Intelligent transport systems, community care and healthcare systems, food production and distribution, and calamity prevention are examples of new research and development investment priority targets.

Mobility is a key to ensuring that individuals can live actively. More security and sustainability are also important in the transport system, so developing driverless cars is just as crucial as developing electric ones.

Table 7.1 compares society development stages for different points of view.

Scharmer [21, 22] discusses the social stages, Helbing [8, 9] involve the economy stages and Keidanren [13] refers to the society stages.

Observe that the points of views are similar. The Keidanren approach uses the labels 1.0 to 5.0 were other initiatives or authors use the labels 0.0 to 4.0 for similar stages.

O. Iordache, *Polytopic Roadmaps*, Studies in Systems, Decision and Control 368 ,
https://doi.org/10.1007/978-3-030-75630-7_7

Table 7.1 Stages comparison

Stage	Social stage	Economy	Keidanren
0.0	Communal	Primitive	Hunting 1.0
1.0	State centric	Agriculture	Agrarian 2.0
2.0	Free market ego-centric	Industry	Industrial 3.0
3.0	Social market regulated	Service	Information 4.0
4.0	Co-creative eco-centric	Digital	Super Smart 5.0

In fact, it is considered that there are five distinct development stages.

The stage 4.0 as described by Scharmer [21] can be compared to the stage 5.0 as described by Keidanren [13].

Observe that comparable five development stages denoted 0.0, 1.0, 2.0, 3.0 and 4.0 appear in the study of society 4.0 but also in the VDMA industry 4.0 toolbox [1].

Recall that dynamic skill theory outlined four sub-stages denoted 0, 1, 2, 3 for any developmental level or tier [5, 6].

Keidanren presented strategy incorporates a human-centered approach.

On this basis, we could characterize polytopic society, in terms of Polytopic Roadmaps.

Selecting for instance four domains, from Table 1.1 each having five stages denoted by 0.0 to 4.0 we arrive at the 6D frameworks for society. For higher dimensional polytopes the term polytopic society is proposed.

Table 7.2 selects from Table 1.1 four domains corresponding to case a.

The selected basic domains have been:

- Nature
- Learning
- Labor
- Economy

Figure 7.1 illustrates society 5.0 case a.

Table 7.2 Four domains-case a

	Nature	Learning	Labor	Economy
0.0	Mother nature	Tradition	Self sufficiency	Tradition
1.0	Resources	Memorization	Slave	Agriculture
2.0	Commodity land materials	Open access library internet	Labor commodity	Industrial
3.0	Regulated commodity	Knowledge producing	Labor regulated commodity	Service
4.0	Eco-system Commons	Innovation producing	Social business entrepreneur	Digital

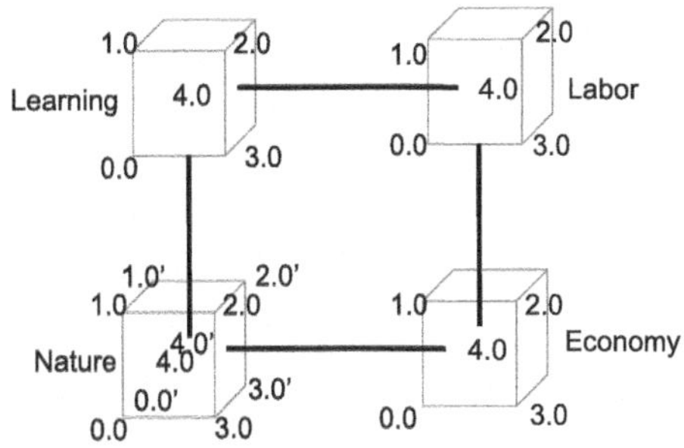

Fig. 7.1 Polytope 6D for society 5.0 case a

This case a, refers to the societal choice in which based on Nature resources, by Learning and Labor the Economy objective is targeted.

The polytope shows how to close the loop Nature to Economy.

Figure 7.2 shows the Polytopic Frameworks for society 5.0 case a.

Figure 7.3 show the polytope 7D for society 5.0 case a.

The strategy option was conservative.

The modified basic domains have been denoted:

- Nature'
- Learning'
- Labor'
- Economy'

Selecting other four domains, from Table 1.1 each having five stages denoted by 0.0 to 4.0 we arrive at another 7D framework for society.

Table 7.3 selects four domains corresponding to case b.

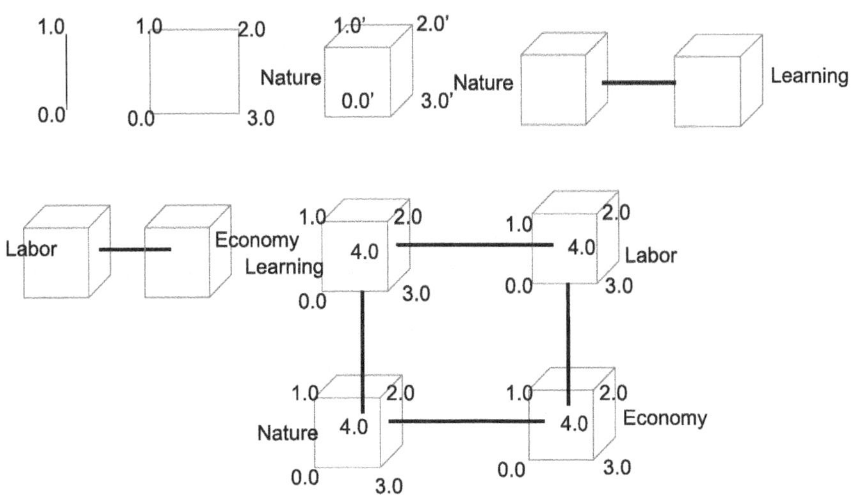

Fig. 7.2 Polytopic roadmaps for society 5.0 case a

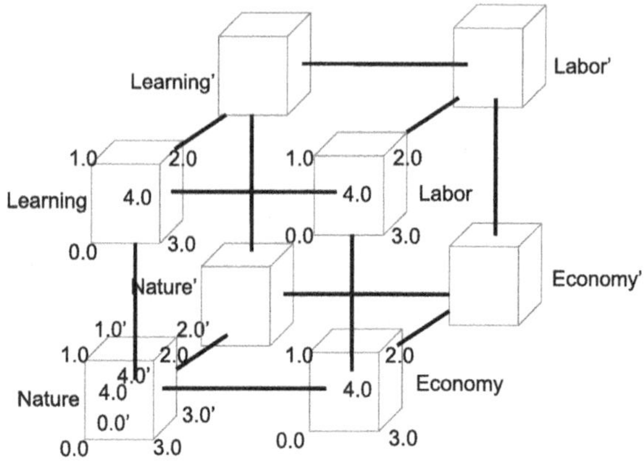

Fig. 7.3 Polytope 7D for society 5.0 case a

Table 7.3 Four domains-case b

	Health	Transport	City	Marketing
0.0	Tradition	Individual	Tradition	Individual
1.0	Examination dialogue	Public transport	Private	Product centric
2.0	Open access Internet	Private cars	City regulated	Customer centric
3.0	Semantic personalized	Autonomous vehicles producing	Private/city regulated	Human centric
4.0	Symbiotic conscious	Uber transport systems	Smart city	Self-Actualization

Figure 7.4 illustrates society 5.0 case b.
The selected basic domains have been:

- Health
- Transport
- Cities

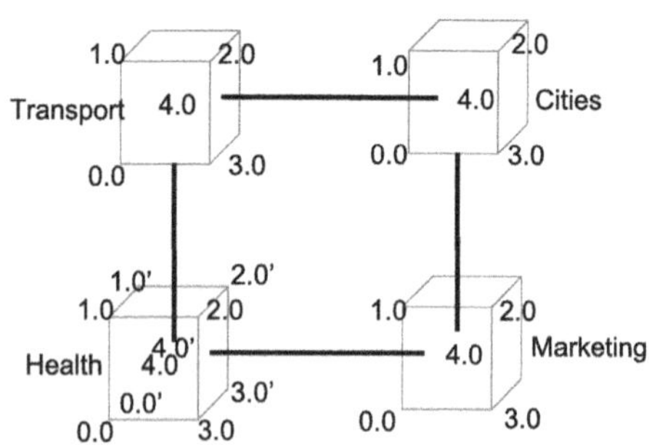

Fig. 7.4 Polytope 6D for society 5.0 case b

- Marketing

This case b refers to the societal choice in which based on Health care, Transport and Cities facilities the advance towards Marketing is targeted.

The polytope shows how to close the loop Health to Marketing.

Observe that there are multiple aspects of polytopic society, according to the interest of different members of the society. The individualized concept of polytopic society depends on age and on the general development of society.

Horizons beyond society 4.0 and society 5.0 towards high-dimensional polytopic society are waiting to be identified and explored.

Figure 7.5 shows the Polytopic Roadmaps for society 5.0 case b.

Figure 7.6 show the polytope 7D for society 5.0 case b.

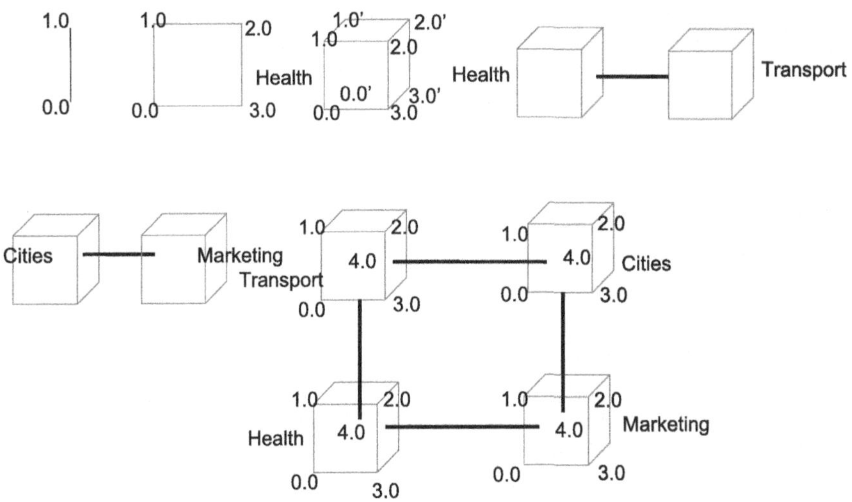

Fig. 7.5 Polytopic roadmaps for society 5.0 case b

Fig. 7.6 Polytope 7D for society 5.0 case b

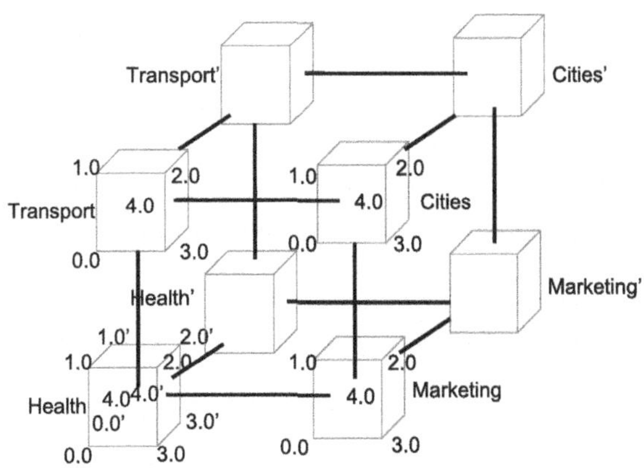

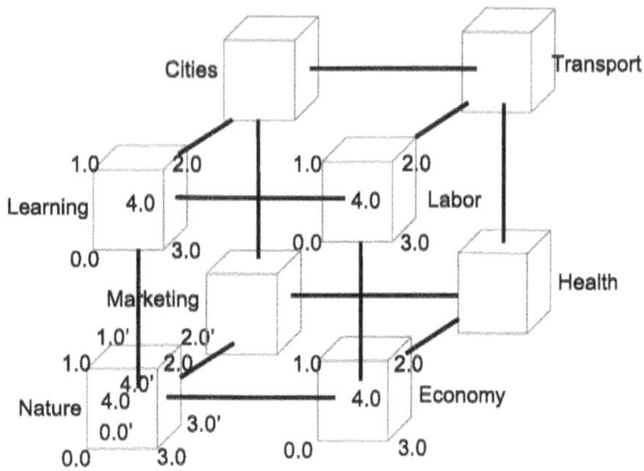

Fig. 7.7 Polytope 7D for society 5.0 case a and case b

We selected a conservative strategy for development.

The modified basic domains have been denoted:

- Health'
- Transport'
- Cities'
- Marketing'

Obviously it is possible to adopt an exploratory approach and avoid restricting to simple modification for basic level or large scale.

For example we can put together the case a, and case b domains shown in Tables 7.2 and 7.3.

Figure 7.7 shows the Polytope 7D for society 5.0 case a, and case b.

The domains Nature, Learning, Labor and Economy refer to case a while the domains Health, Transport, Cities and Marketing refer to the case b.

Together they allow closing the loop from Nature to Marketing.

Observe that exploratory strategy needs eight domains while the conservatory strategy needs four domains.

Figure 7.8 shows a Periodic Table for polytopic society.

The Periodic Table contains columns called groups and rows called periods.

The columns correspond in succession to dots, lines, squares, cubes and hypercubes that is to 0D, 1D, 2D, 3D and 4D. The columns highlight the dot, line, square, cube and hypercube pattern.

The first row use as basic bricks the 0D (dots), the second row use 4D (hypercubes) and the third row use 8D (hypercube of hypercubes) .

The nD frame is associated to society n.0 or in other words to polytopic society.

Notice that we considered the society 4.0 as the so-called society everything 4.0.

The Periodic Table contains the polytopes 6D and 7D as resulting from society 5.0 analysis.

This table may help to illustrate the geometry of frameworks naturally associated to different stages of society.

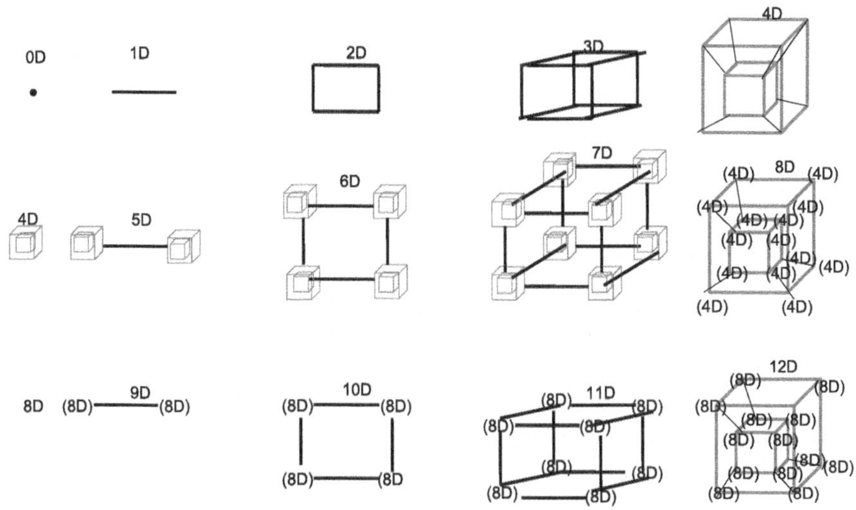

Fig. 7.8 Periodic Table for polytopic society

The next row unincluded in the Periodic Table from Fig. 7.8 will show 12D, 13D, 14D, 15D and 16D frames.

7.2 BINC

The polytopic projects joint recent trends advocating the convergence of several disciplines as, nanoscience, biotechnology, information technology and cognitive science known as the NBIC concept [3].

This concept was developed also under the label BINC (bio-info-nano-cogno convergence) [20].

Convergence is a new paradigm that can yield critical advances in a broad array of sectors, from health care to energy, food, climate and other domains of interest for society [23].

It should be emphasized that the convergence way does not quite grasp the essence of inventiveness and creativity required to confront high complexity problems. Tendencies to converge will be correlated with tendencies to diverge and it is the skillful relation of both trends in a polytopic architecture that matters. This is significant for technologies, data analysis, artifacts and for theories.

BINC approach may be seen also as an implementation of self-integrative closure concept were the convergence is referring to matter in general and is not restricted to nano domain [10–12].

The idea for integrative closure was to reassess and completes the interrelation between the four basic ontological levels in the study of nature: material, biological, cognitive or psychological and intelligent or logical [7].

It was acknowledged that the hierarchical structures cannot serve as unique model for multi-level knowledge organization. Facing complexity the task of knowledge integration remains pertinent. Integrative closure is the strategy allowing confronting complexity.

The four basic levels of real systems have been denoted: S (or K0), K1, K2 and K3.

The integrative closure aims to make ends meets, for the four levels or realms, emphasizing the hypothetical interconnection between the material realm, S and intelligent realm, K3. The closed system is supposed to cross the gap between these farthest levels. This gap crossing corresponds to the cyber-physical systems. The cyber part is associated to K3 while the physical part is associated to S.

Integrative closure hypothesis is looking for structural analogies and a common methodology for different domains as matter-bio-cognitive-intelligent inspired systems, knowledge organization, problem solving or technological developments.

The growth in complexity is not only about differentiation but also about integration and coordination. The pursuit of evolvability towards further step of self-evolvability for systems imposed the study of polytopic architecture associated to the general framework as shown in Fig. 7.9. Figure 7.9 illustrates the concept of self-integrative closure with specific reference to the four levels of the reality.

The basic levels are S, K1, K2 and K3 represented on the front face of the outer cube and S′, K1′, K2′ and K3′ represented on the back face of the outer cube of the 4D polytope.

Following Hartmann ontology, for the polytope shown in Fig. 7.9 we could identify:

- S-Nano
- K1-Bio
- K2-Cogno
- K3-Info

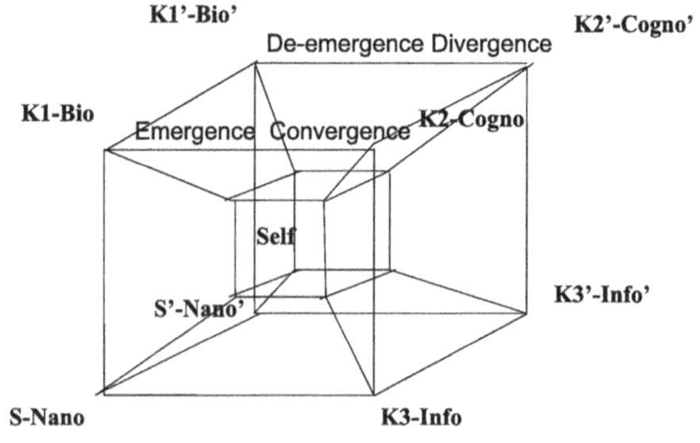

Fig. 7.9 Self-integrative closure for BINC

Figure 7.9 shows the self-integrative closure for BINC.

For engineering systems we may refer to a hierarchy of artificial systems inspired by the four level of Hartman's ontology as for instance:

- S-Materials or Smart Materials, Self-Evolving or Living Technologies
- K1-Bio-Inspired or Biomimetics Systems, Artificial Life
- K2-Cognitive-Inspired Systems, Neural Computer, Artificial Brain
- K3-Artificial Intelligent System, Cyber-Physical Systems

The objects of engineering are manufactured articles or artificial systems. The reality levels as material biological, cognitive and intelligent may serve for inspiration purposes for engineering.

This is an example of reality inspired engineering (RIE).

The swinging between the two faces of the outer large cube in the polytope shown in Fig. 7.9 is mediated by the inner small cube identified also as "Self".

The polytope outlines two sequences for evolution the direct sequence S → K1 → K2 → K3 that corresponds to the emergence stages and the reverse sequence K3′ → K2′ → K1′ → S′ that corresponds to the modified stages.

This corresponds to a conventional or conservative type of development. The acquired development of the direct sequence is questioned and modified by the reverse sequence.

This reference framework was labeled as self-integrative closure [10, 11].

Modifications of the direct sequence S, K1, K2, K3 during reverse sequence K3′, K2′, K1′ and S′ may happen and be accepted.

We may identify:

- S′-Nano′
- K1′-Bio′
- K2′-Cogno′
- K3′-Info′

Emergence or convergence from the direct sequence is complemented by de-emergence, or divergence in the reverse sequence. During these complementary processes of emergence, and de-emergence, the differences allowing and accompanying self-evolvability and smartness may be generated. The terms convergence and divergence have been also used for direct and reverse sequence.

The fourth industrial revolution was considered as a more radical BINC [20].

The first industrial revolution was the mechanization of production using water, wind and steam power (industry 1.0) .

During the nineteenth century, the industrial revolution developed mass production in factories and a vast transportation infrastructure (industry 2.0). In the latter part of the twentieth century and the start of current century, the information technology revolution programmed personal information processing in computers (industry 3.0) . The next major technological revolution will be based on an integration of information processing and material production akin to living organisms and ecosystems combining these processes (industry 4.0) .

Major transitions in human societies are usually characterized by new tools and new infrastructures, new ways of making a living, changes in trade, new institutions and organizational forms, new local and global power structures as well as a new collective narrative [20]. The "Self" of polytopic projects may be associated to narratives of BINC.

Globalization, mainly powered by information technology, is presenting us with a complex network of interconnected phenomena, including taxation and control issues for global companies, profit from digital products without production, which together with other issues cause the nation state to lose its influence over its own destiny. Transparency and privacy issues are increasingly taking the spotlight, because the new technologies generate new questions about who should have insight in what, and who own or has access to an individual's data.

The BINC interface will soon enable individuals to design, transmit, produce and recycle most needed material goods as well as personalized medicine within the confines of their home through personal fabrication devices. This represents interesting opportunities for advances in regional sustainability as well as major challenges for current concepts in economy, future employment and ownership of the means of production.

The new opportunities and challenges we face through the BINC technology convergence require that all components of society engage with these new technologies trends in a responsible manner and work together to develop a free and open society.

Figure 7.10 shows the Polytopic Roadmaps for BINC.

The basic realms or levels are denoted: Nano, Bio, Cogno and Info.

Within each realm there is a recurring cycle of four sub-levels:

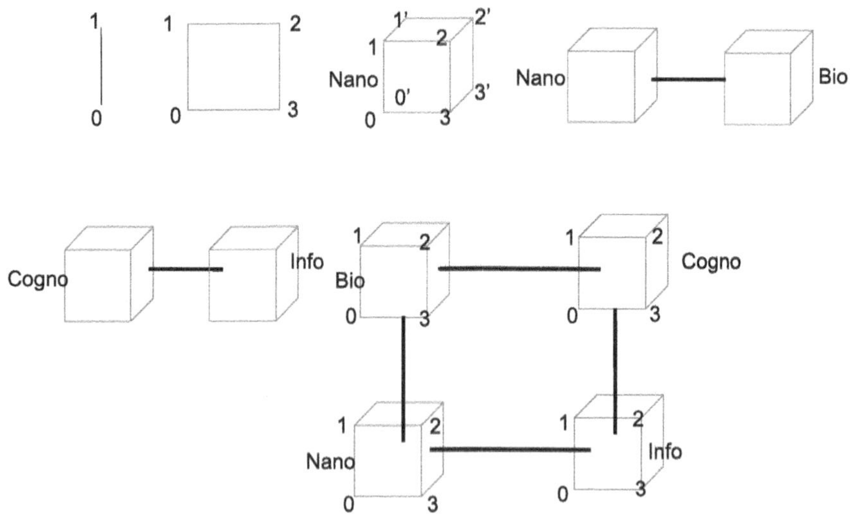

Fig. 7.10 Polytopic roadmaps for BINC

0-single set
1-mapping (of sets)
2-system (composition of mappings)
3-system of systems

The decomposition in sub-levels as: set-0, mapping-1, system-2 and system of systems-3 follow suggestions from dynamic skill theory of development [5].

Notations as 0′, 1′, 2′, 3′ corresponds to modifications or improvements.

0′-single set'
1′-mapping′
2′-system'
3′-system of systems'

As shown in Fig. 7.10 the Polytopic Roadmaps results by doubling and relating.

It is supposed that there is an evolution toward the Nano level and from this towards the Bio level. The preliminary steps are 0D, 1D and 2D. The basic levels are 3D.

A similar evolution involves the Cogno and Info levels before relating the four realms in BINC.

Figure 7.11 shows the polytope 5D for BINC.

Using the thick hyper-lines, the 5D cube is shown as square of cubes.

Figure 7.12 shows the polytope 6D for BINC.

With the thick hyper-lines, the 6D cube is shown as cube of cubes.

We selected a conservative strategy for development.

The basic realms are: Nano, Bio, Cogno and Info are modified to Nano', Bio', Cogno' and Info'.

This modification may be due to a modified strategy. This new strategy plays the central role in transition from 5D to 6D.

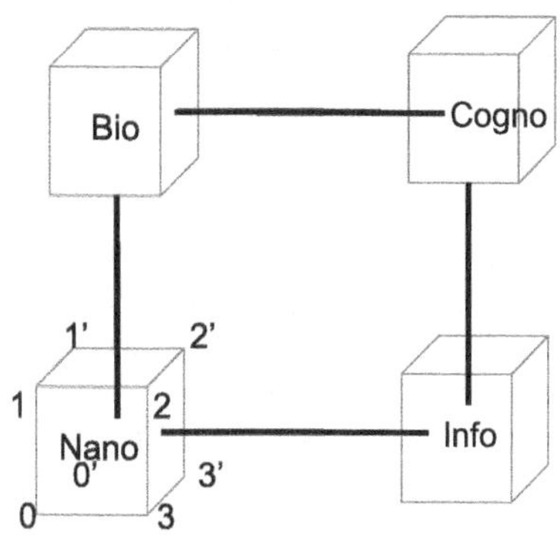

Fig. 7.11 Polytope 5D for BINC

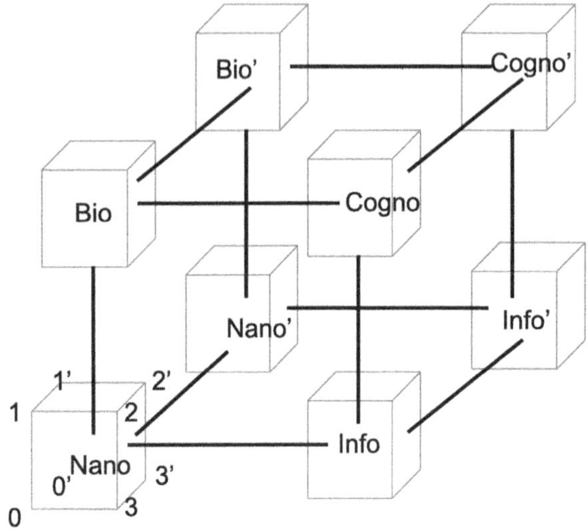

Fig. 7.12 Polytope 6D for BINC

7.3 Cyber-Physical-Social Systems

Cyber-physical systems (CPS) are smart systems that have cyber technologies, both hardware and software, deeply embedded in and interacting with physical components, sensing and changing the state of the real world. They represent a core opportunity area and source of competitive advantage for the innovation economy.

Examples of CPS-based initiatives include, smart factory, smart transportation systems, smart medical devices and technologies, smart buildings, next-generation air transportation systems, and the Smart Grid [2, 4].

For CPS the real (physical) objects and processes associated with information processing (virtual) objects and processes are anytime interconnected information networks. Optionally a CPS may have human-machine interfaces and provides the ability to dynamically adjust the system [27].

The term cyber-physical systems (CPS) characterizes the new generation of systems with integrated computational and physical capabilities that can interact with humans through many new modalities. The ability to interact with, and expand the capabilities of, the physical world through computation, communication, and control is the key enabler for future technology developments. Opportunities and research challenges include the design and development of next-generation devices, cars, airplanes, hybrid gas-electric vehicles, fully autonomous urban driving or factories, and prostheses that allow brain signals to control physical objects [16, 24].

CPS research aims to integrate knowledge and engineering principles across the computational and engineering disciplines (networking, control, software, human interaction, learning theory, as well as electrical, mechanical, chemical, biomedical, material science, metallurgical and other to develop new CPS science and supporting technology.

Many engineering systems have been conceived by decoupling the control system design from the hardware and software implementation details. After the control system is designed and verified by extensive simulation, unstructured tuning methods have been used to address modeling uncertainty and random disturbances. The integration of various subsystems, while keeping the system functional and operational, has been time-consuming and costly.

The increasing complexity of components and the use of more advanced technologies for sensors and actuators, wireless communication, and multi-core processors pose a major challenge for building next-generation device control systems. Both the supplier and integrator need new systems science that enables reliable and cost-effective integration of independently developed system components. In particular, theory and tools are needed for developing cost-effective methods to:

- Design, analyze, and verify components at various levels of abstraction, including the system and software architecture levels, subject to constraints from other levels
- Analyze and understand interactions between the device control systems and other subsystems
- Ensure safety, stability, and performance while minimizing device cost
- The CPS systems have been introduced to confront such challenges.

The CPS allows the fusion of computation and a physical process. Typically, a CPS is composed of physical process monitored and controlled by a cyber system, which is a networked system of distributed sensing, communication, and computational devices.

In a CPS, heterogeneous embedded systems monitor and control a physical process, usually via feedback loops where the physical process affects computation and vice versa. Emerging CPS devices are characterized by large-scale system sizes, heterogeneity of resources, uncertain system dynamics, and extensive physical interactions.

Unlike traditional embedded systems aiming to optimize computation in highly constrained environments using limited resources, the emphasis of CPS research tends to be more on tight coupling between the computing and physical parts. Embedded systems can be considered CPS when they sense or influence dynamic physical environments.

CPS may be considered as part of the next computing revolution and has been identified as a new key area of research and engineering. CPS applications include high-confidence control and safety, advanced automatic systems, energy conservation, environmental control, instrumentation, critical infrastructure control, defense systems, distributed robotics, industrial control systems, smart structures, cellular communication, and autonomic computing systems.

CPS design is an interdisciplinary research area of computer architecture, software, network, physical systems, control, and other engineering disciplines. This emerging area enables new opportunities and poses additional research challenges including CPS decomposition and interoperability, robustness safety and security, control of hybrid systems, real-time embedded system abstractions, architecture development, and model-based development.

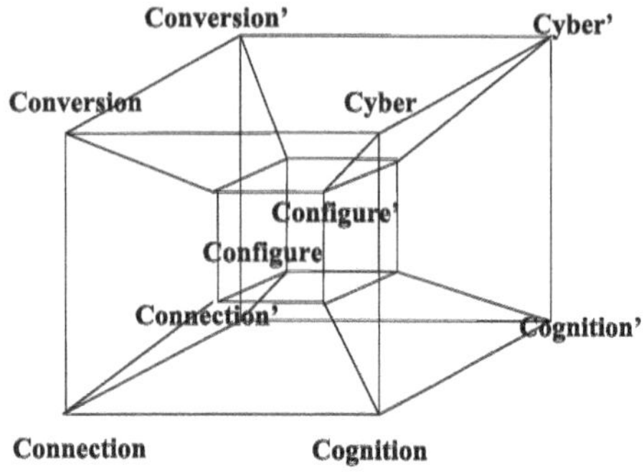

Fig. 7.13 Cyber-physical systems-5 levels architecture

Typically, a CPS is configured as a closed-loop feedback control system.

Figure 7.13 illustrates a basic CPS architecture with 5-levels.

These may be associated to the polytopic architecture as follows [16]:

- S-Connection
- K1-Conversion
- K2-Cyber
- K3-Cogniton
- K4-Configure

Embedded controllers are computational systems incorporating a set of components such as processors, memory, storage, and input and output devices. This infrastructure is abstracted to layers in order to simplify the development flow. System abstraction layers include hardware, operating system, software, and data. Modern components are characterized by the complexity of functions and interactions, with data traversing through different abstraction layers.

We selected a conservative strategy for the polytope shown in Fig. 7.13.

The 5th level is the center able to control the configuration.

Implementing polytope project for cyber-physical systems starts from the direct sequence S → K1 → K2 → K3 and complements this by the reverse sequence: K3′ → K2′ → K1′ → S′. For the polytope shown in Fig. 7.13 we can identify modifications as follows:

- K4′-Configure'
- K3′-Cognition'
- K2′-Cyber'
- K1′-Conversion'
- S′-Connection'

This means to put together the cyberizing of the physical and the physicalizing of the cyber. These two complementary approaches have been identified and studied by Lee [14, 15]. Cyberizing the physical, from S to K3, means to endow physical subsystems with cyber-like abstractions and interfaces. Physicalizing the cyber, from K3′ to S′ means to endow software and network components with abstractions and interfaces that represent their dynamics in time.

The CPS architecture based on 5-levels was associated to industry 4.0 manufacturing systems [16]. This architecture was expanded by Yin et al. [26] to 7-level architecture. The aim was to include social elements in the CPS.

Social elements are new levels.

In this way we will look at CPSS that is Cyber-Physical-Social System.

A frame with 8-levels is proposed here. The 8th level corresponds to the understanding.

This happens after the collective intelligence manifestation.

The 8-levels including the 7-levels investigated by Yin et al. [26] are:

- S-Connection
- K1-Conversion
- K2-Communication
- K3-Computing
- K4-Cogniton
- K5-Configuration
- K6-Collective intelligence
- K7-Understanding

Figure 7.14 shows the cyber-physical-social-systems-8 levels architecture.

For the CPSS shown in Fig. 7.14 we selected an exploratory innovative strategy.

This explains why we need 8-levels.

A cyber-physical-social system considers human factors as a part of a system instead of placing them outside the system boundary [25].

The CPSS provide an ideal paradigm for the design and construction of command and control organizations [17].

A framework for command and control is shown if Fig. 7.15.

The main stages of organization and their abbreviations are:

- S-Physical
- K1-Cognitive
- K2-Organizational
- K3-Cyber

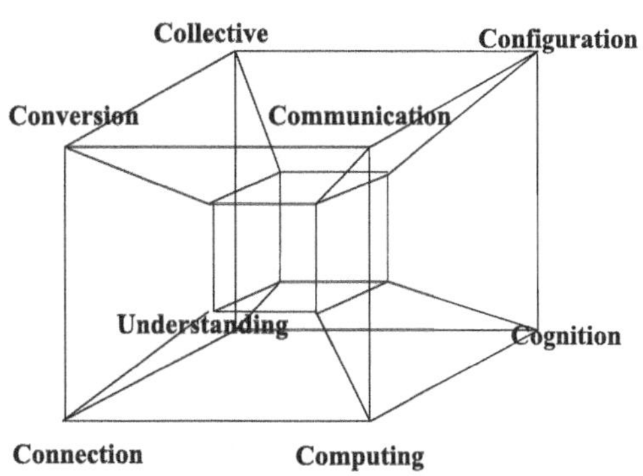

Fig. 7.14 Cyber-physical-social systems-8 levels architecture

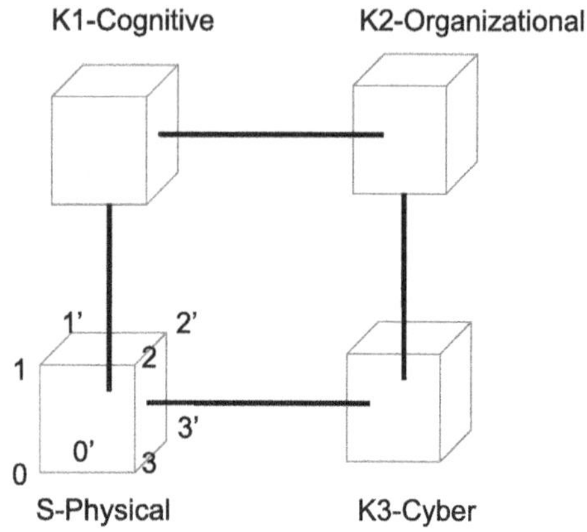

Fig. 7.15 Framework for command and control

S corresponds to artifacts networks, K1 to a cognitive network, K2 to an organizational coordination network and K3 to the cyber network.

Sub-stages may be considered for any of these four stages.

For instance for the stage S we may observe the sub-stages [17]:

0-devices, 1-sensors, 2-enablers, 3-communications

These parallel the sequence 0-singles, 1-mappings, 2-systems, 3-SoS.

The sub-stages 0′, 1′, 2′, 3′ denotes modifications or improvements.

Figure 7.16 shows the polytope 6D for developed operational mechanism including modifications and improvements.

For the polytope shown in Fig. 7.16 we selected a conservative strategy for modifications.

Physical', Cognitive', Organizational', and Cyber' denotes the modified stages.

The modification may be operated for the stage or for the sub-stage.

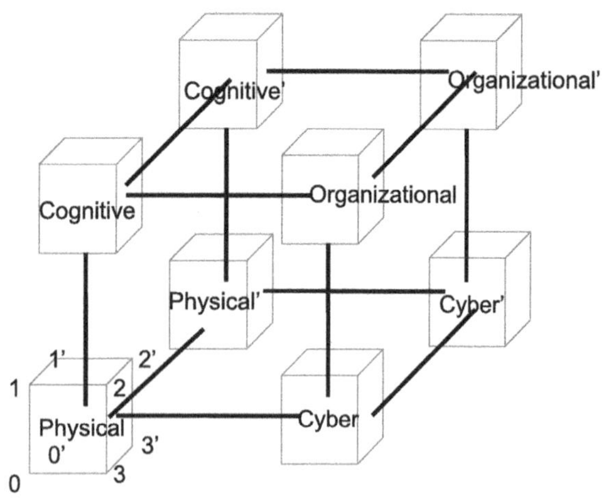

Fig. 7.16 Polytope 6D for command and control

7.4 Ecology

Polytopic modeling architectures may be useful in ecological studies dedicated to deep ecology and to permaculture [18, 19].

Deep ecology is an ecological philosophy distinguished by its advocacy of the inherent worth of living beings regardless of their instrumental utility to our needs. Deep ecology argues that the natural world is a subtle balance of complex interrelationships in which the existence of organisms is dependent on the existence of others within ecosystems.

Interference with or destruction of the natural world poses a threat therefore to all organisms constituting the natural order.

The prevailing attitude in deep ecology is that an environment-oriented activity should not be performed passively but rather proactively for the goal of self-realization. This takes the "Self" as being woven into the whole world of entities and events without temporal limits of past and future and without spatial limitations.

Figure 7.17 shows the deep ecology diagram.

The polytope outlined two sequences for evolution the direct sequence S → K1 → K2 → K3 that corresponds to the main development stages and the reverse sequence K3′ → K2′ → K1′ → S′ that corresponds to the modified stages.

This is a conventional or conservative type of development. The derivative development is questioned and modified.

The notations are:

- S-Premises
- K1-Platforms
- K2-Norms and Values
- K3-Actions

The modified levels are denoted:

- S'-Premises'
- K1′-Platforms'
- K2′-Norms and Values'
- K3′-Actions'

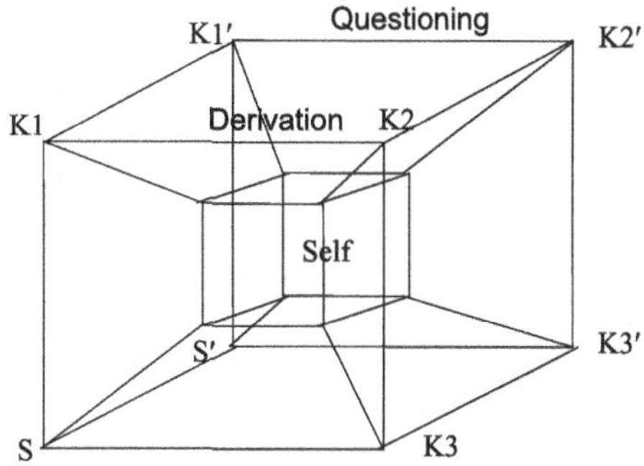

Fig. 7.17 Deep ecology diagram

The polytope outlines the direct way S → K1 → K2 → K3 from premises to conclusions based on logical derivation. Then we need to look at the indirect way K3′ → K2′ → K1′ → S′ based on questioning.

Schematically, we may represent the total views implied in the movement by streams of derivation from the most fundamental norms and descriptive assumptions to particular decisions in actual life situations (Fig. 7.17).

The basic endeavor of our study is to outline and implement the isomorphism or parallelism between the cognitive-intelligent development stages, and the smart operations, equipments, technology and industry development stages. It is the general isomorphism with respect to the higher dimension development and cyclic control mechanisms.

This isomorphism explains why the ecological model illustrated by Fig. 7.17 has significant features in common with hypothetical-deductive systems.

The main difference, however, is that some sentences at the top or deepest level are normative, and are preferably expressed by imperatives. This makes it possible to arrive at imperatives at the lowest derivational level, the critical level in terms of decisions.

Just as in a hypothetical-deductive system in physics, where only the two upper levels of the direct way are thought of as forming physics as a system, so also in normative systems only the upper levels are considered to be part of the total system.

The sentences in the lowest part may change from day to day as life situations change.

After derivation and direct way we look at the indirect way K3′ → K2′ → K1′ → S′.

The indirect way is a questioning based on philosophical reflection.

The deep ecology movement questions the presuppositions of shallow argumentation. Even what counts as a rational decision is challenged, because rational is always defined in relation to specific aims and goals. If a decision is rational in relation to the lower-level aims and goals of our direct or indirect ways but not in relation to the highest level, then the decision should not be judged to be just rational. If an environmentally oriented policy decision is not linked to intrinsic values, its rationality is yet undetermined. The deep movement connects rationality with a set of philosophical foundations. One cannot expect the ultimate premises to constitute rational conclusions. There are no deeper premises available.

The deep ecological questioning reveals the fundamental normative orientations. Shallow argumentation stops before reaching fundamentals or jumps from the ultimate to the particular, that is, from level S to level K3.

It is not only normative claims that are in the balance. Most norms presuppose ideas about how the world functions. Typically, the vast majority of propositions needed in normative systems are descriptive. This holds of all levels.

Permaculture is that branch of ecological design, ecological engineering, and environmental design which develops sustainable framework and self-maintained horticultural systems modeled from natural ecosystems. Permaculture design emphasizes patterns of landscape, function, and species assemblies. The central concept of permaculture is maximizing useful connections between components and synergy of the final design. The focus of permaculture is not on each separate element, but rather on the relationships created among elements by the way they are placed together.

Permaculture design seeks to minimize waste, labor, and energy input by building systems with maximal benefits between design elements to achieve a high level of synergy. Permaculture designs evolve over time by taking into account these relationships and elements and can become highly complex systems that produce a high density of food and materials with minimal input.

The design principles which are the conceptual foundation of permaculture were derived from the science of systems ecology and study of examples of sustainable land use.

The edge effect in ecology is the effect of the juxtaposition or placing side by side of contrasting environments on an ecosystem. Permaculture developers argue that, where vastly differing systems meet, there is an intense area of productivity and useful connections. Edges between woodland and open areas have been claimed to be the most productive.

References

1. Anderl, R, Picard, A., Wang, Y., Fleischer, J., Dosch, S., Klee, B., Bauer, J.: Guideline industry 4.0—guiding principles for the implementation of Industry 4.0 in small and medium sized businesses. In: VDMA Forum Industry 4.0, Frankfurt (2015)
2. Baheti, R., Gill, H.: Cyber-physical systems. Impact Control Technol. **12**, 161–166 (2011)
3. Bainbridge, W.S., Roco, M.C.: Managing Nano-Bio-Info-Cogno Innovations: Converging Technologies in Society. Springer, Berlin (2006)
4. Fang, X., S. Misra, S., Xue, G. Yang, D.: Smart grid, the new and improved power grid: a survey. IEEE Commun. Surv. Tutorials **14**(4), 944–980 (2012)
5. Fischer, K.W.: Dynamic cycles of cognitive and brain development: measuring growth in mind, brain, and education. Educ. Brain Essays Neuroeducation 127–150 (2008)
6. Halford, G.S., Andrews, G.: Information-Processing Models of Cognitive Development. The Wiley-Blackwell Handbook of Childhood Cognitive Development. **16**, 697–722 (2010)
7. Hartmann, N.: The New Ways of Ontology. Greenwood Press, Westport (1952)
8. Helbing, D.: What the digital revolution means to us. Sci. Bus. (2014, June 12)
9. Helbing, D.: Societal, economic, ethical and legal challenges of the digital revolution: from big data to deep learning, artificial intelligence, and manipulative technologies. Artificial Intelligence, and Manipulative Technologies (2015, April 14)
10. Iordache, O.: Self-Evolvable Systems. Machine Learning in Social Media. Springer, Berlin, Heidelberg (2012)
11. Iordache, O.: Polytope Projects. Taylor & Francis, CRC Press, Boca Raton, USA (2013)
12. Iordache, O.: Implementing Polytope Projects for Smart Systems. Springer, Cham, Switzerland (2017)
13. Keidanren: Reform of the Economy and Society by the Deepening of "Society 5.0". Keidanren Japan Business Federation (2016)
14. Lee, E.A.: Computing foundations and practice for cyber-physical systems: a preliminary report. Tech Report University of California Berkeley/EECS-2007-72 (2007)
15. Lee, E.A.: CPS foundation. In: Proceedings of the 47th Design Automation Conference, Anaheim, CA, pp. 737–742 (2010)
16. Lee, J., Bagheri, B., Kao, H.A.: A cyber-physical systems architecture for industry 4.0-based manufacturing systems. Manuf. Lett. **3**, 18–23 (2015)

17. Liu, Z., Yang, D.S., Wen, D., Zhang, W.M., Mao, W.: Cyber-physical-social systems for command and control. IEEE Intell. Syst. **26**(4), 92–96 (2011)
18. Mollison, B., Slay, R.M.: Introduction to Permaculture. Tagari Publications, Australia (1991)
19. Næss, A.: Ecology, Community and Lifestyle: Outline of an Ecosophy. Cambridge University Press, Cambridge, Cambridge (1989)
20. Rasmussen, S.: The "4th Industrial Revolution" in fact a more radical "BINC Revolution"? In: Conference American Chamber of Commerce, China, Beijing (2016, December 14)
21. Scharmer, O.: Theory U: Leading from the Future as It Emerges. Berrett-Koehler, San Francisco (2009)
22. Scharmer, O., Kaufer, K.: Leading from the Emerging Future: From Ego-System to Eco-System Economies. Berrett-Koehler, San Francisco (2013)
23. Sharp, P.A., Langer, R.: Promoting convergence in biomedical science. Science **333**, 527 (2011)
24. Strang, D., Anderl, R.: Assembly process driven component data model in cyber-physical production systems. World Congr. Eng. Comput. Sci. **2**, 22–24 (2014)
25. Wang, F.Y.: The emergence of intelligent enterprises: From CPS to CPSS. IEEE Intell. Syst. **25**(4), 85–88 (2010)
26. Yin, D., Ming, X., Zhang, X.: Understanding Data-Driven Cyber-Physical-Social System (D-CPSS) using a 7C framework in social manufacturing context. Sensors **20**, 5319 (2020)
27. Yuan, X., Anumba, C.J., Parfitt, K.M.: Review of the potential for a cyber-physical system approach to temporary structures monitoring. Int. J. Archit. Res. ArchNet-(IJAR) **9**(3), 26–44 (2015)

Chapter 8
Perspectives

8.1 Integral Perspectives

8.1.1 General Systems Theory

General Systems Theory (GST) raises from several disciplines, including physics, biology, cognitive sciences, social sciences, philosophy, and mathematics.

The goal of GST was to find isomorphic structures in different disciplines, to uncover a method for reducing replication of experimental and theoretical effort in the sciences [27, 31].

It was expected that GST could serve as a unifying theoretical construction for different sciences. To this end, scientists and engineers became engaged in creating integration theories and applications.

The theory arose out of the efforts of researchers who were looking for isomorphism or parallelism between works in the physical, biological, cognitive and social sciences.

General Systems Theory in the broadest sense refers to a collection of general concepts, principles, tools, problems, methods, and techniques associated with systems [19]. A system is an arrangement of certain components so interrelated as to form a whole. The systems approach arose in contrast to the Cartesian splitting tradition of separating an object into its component parts and attempting to understand the behavior of the object by understanding the properties of the individual parts while overlooking their interactions.

Compliant with Leibnitz's relational world philosophy, in systems studies, the reality is considered as the totality of interactions, as a whole.

GST is a set of related definitions, assumptions, and propositions which deal with reality as an integrated hierarchy of organizations of matter and energy [23]. It is particularly applicable to systems with goal seeking behavior. Biological cognitive and social systems are the primary domains of such goal seeking behavior.

Major functions for GST have been:

O. Iordache, *Polytopic Roadmaps*, Studies in Systems, Decision and Control 368 ,
https://doi.org/10.1007/978-3-030-75630-7_8

- Investigate the isomorphism or parallelism of concepts, laws, and models in various fields
- Help in useful transfers from one field to another
- Promote the development of adequate theoretical models in the fields which lack them
- Minimize the replication of theoretical and practical effort in different fields
- Endorse the unity of sciences through improving communication among specialists of different disciplines [31].

Criticisms of GST included the concern that this theory was irrelevant, false and misleading, or unsafe methodologically and philosophically.

The concern over the irrelevant nature of GST arose because some critics thought GST only provided insights analogous to the idea that the same model could be used to describe phenomena in multiple fields. In fact, the aim of GST was not to find a model but to make available a formal logical structure able to describe the relationship between different parts of a system.

The concern over the false and misleading nature of GST arose because some critics thought the theory made false analogies such as equating fluid systems with organisms or evolving societies. Superficial analogies can't justify unity of mechanisms and theories.

The concern over the philosophical and methodological misleading arose because researchers in individual fields were concerned that GST researchers would be diverted from the real problems within their field of expertise.

GST was correlated to cybernetics, information theory, game theory, decision theory, systems theory, topology, and factor analysis as complementary scientific domains. Areas of engineering application include systems engineering, operations research, economy and ergonomics.

In the sense of fulfilling the proposed goals, GST has partially succeeded [27].

It has not succeeded in finding the isomorphism or parallelism of concepts, laws, and models in various fields, in providing a unifying framework for different sciences. Additionally it was difficult to measure whether researchers in different fields are replicating each other's work less after GST principles implementation then before.

Interconnectivity and awareness of the interrelatedness of everything to everything else is the most important practical result of GST. Ultimately GST succeeded as support of a post-Cartesian relational perspective.

Without doubt, GST will be used by specialized-generalists, that is, peoples who are specialists for a discipline but who also recognize the value and apply the formalism of GST. The usual education starts with specialization and continues with generalization necessary to face high complexity advent. Complexity requires disciplinary education and GST but this road is not enough to face higher complexity. We need both convergent and divergent approaches.

8.1.2 General Polytopic Roadmaps

Essential ways of conveying knowledge should be developed and implemented as the amount of knowledge keeps growing exponentially and the subject areas we deal with are getting exceedingly complicated and challenging.

Hermann Hesse's novel, *The Glass Bead Game* suggests a symbolic essential way to confront high complexity in the twenty-first century [14].

Hesse envisages the glass bead game, (GBG), as a system that has been able to combine technologies, sciences, philosophy and mathematics into one highly condensed expression, a new language with a new vocabulary and grammar.

Projected in a future period, one in which intelligent activity reached its broad expression the GBG provided a way for researchers and engineers from various disciplines to synthesize their thinking into new stages of knowledge [4].

The GBG imagined by Hermann Hesse, look like a high-dimensional abacus, or in modern words a computer or a self-reconfigurable automaton, with several dozen wires strung vertically, horizontally or along several dimensions. Upon these wires were hung beads of various colors, sizes and shapes. Moving the beads into new configurations symbolically represented the development of new themes. Over time, the GBG was adopted by all major disciplines with the beads corresponding to the artifacts, symbols, formulae and notations of that discipline.

As the GBG developed over many years, it became increasingly desirable to develop the common language and grammar across several fields in order to make their similarities and differences clearer and to build an institution around the development, preservation and communication of this new language and paradigm of global culture.

The GBG would transcend different disciplines and allow researchers from these disciplines to interact, and to learn from one another.

The same capability to build upon each other's ideas is described in the current studies of higher complexity. Such studies, concerns the transdisciplinary and selfdisciplinary work in the fields of technology, biology, economics, information sciences and physics, and describes how new insights, for example in market study, emerge from thinking in the field of chemistry or psychology.

In the context of mathematics, the beads of GBG corresponded to mathematical formulae and theorems, which were combined with the mathematical notations of other players, to form new insights. The aim of GBG was to make available mathematical structures able to describe the relationship between different mathematical domains.

In higher dimensional complexity studies workers encounter a ladder which they are irresistibly drawn to ascend, step by step. This ladder proves to be a high dimensional polytope, were the ascending portions are tied to descending ones.

General Polytopic Roadmaps, GPTR, are cognitive-intelligent inspired multipurpose guidelines, useful for artifacts building, information representation, designs, operations and calculus that may be considered as GBG prototypes in action.

GPTR as encountered at different levels of reality and in different applications domains will be presented in what follows in order to emphasize the implemented isomorphism of the frameworks and of control mechanisms.

The main objective of GPTR is to understand and make use of development stage theories as suggestions for new operations, equipments, technologies, industries and societies.

To re-apply this understanding in developing new ways of study or explanations of material-bio-cognitive-intelligent relevance for reality development systems may be considered as a long-term objective only. For instance, self-evolvable computation programs and self-evolvable devices may make use of cognitive-intelligent principles but do not attempt to model or to mimic detailed data or processes from real biosystems. Bio-inspired engineering design is not constrained by high fidelity to the original complex biosystem. It is just dual to real field design.

GST try to find the isomorphism of concepts, laws, and models in various fields, to provide a unifying framework for different sciences of nature while GPTR is an attempt to impose, to built an accepted isomorphism and that unity of different technologies, industries and societies.

It is the difference between the scientific bottom-up approach and the engineering top-down approach. These approaches should work together in duality. GST discovers while GPTR imposes.

GPTR implements his method as for instance generic models that is use of new fields for the same real formal model, to obtain artificial new materials, living, cognitive and intelligent systems.

Some Polytopic Roadmaps will be reviewed in order to highlight their basic similarity despite the marked field difference.

Figure 8.1 shows the General Polytopic Roadmaps. These have been obtained by doubling and relating. These methods increase the dimension from nD to (n + 1) D.

Figure 8.2 shows the general Periodic Table. It is the reorganization in columns and rows of the frameworks shown in Fig. 8.1. The columns correspond to dots, lines, squares and cubes.

Figure 8.3 shows Polytopic Roadmaps for dual graded graphs, DGG. U and D correspond to upward and downward graph. The selected example was integer partition but the roadmap is generally valid for any pair of dual graded graphs.

The 6D framework is the highest dimension targeted in this case.

Figure 8.4 shows the Periodic Table for dual graded graphs.

It reorganizes the frameworks shown in Fig. 8.3.

A linear that is 1D partition completes the framework and is included in the table.

The first row of this Periodic Table shows the elements, 1D, 2D and 3D.

The second row contains the elements 4D, 5D, and 6D.

Both Figs. 8.3 and 8.4 illustrate a continuation and generalization of the dual graded graphs procedures. The versatility of the framework is based on graphs interactions.

Figure 8.5 shows Polytopic Roadmaps for classification. The target is a 5D polytope.

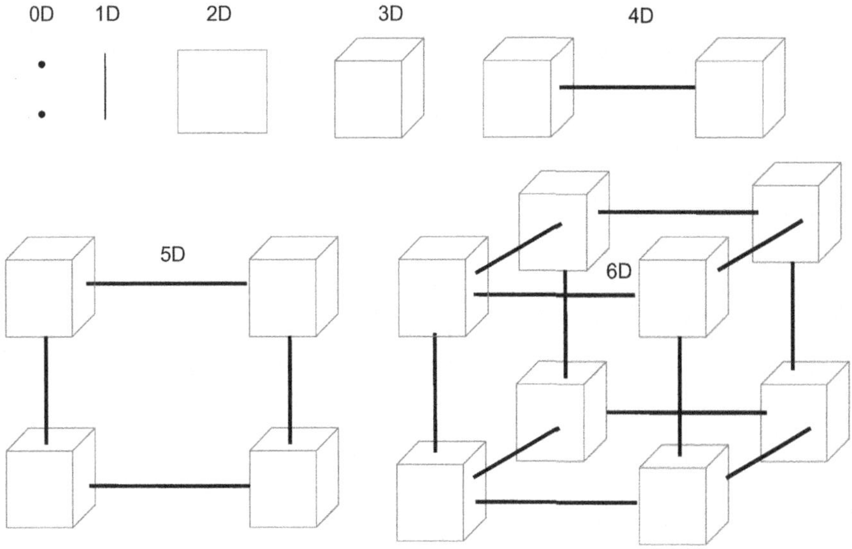

Fig. 8.1 General Polytopic Roadmaps

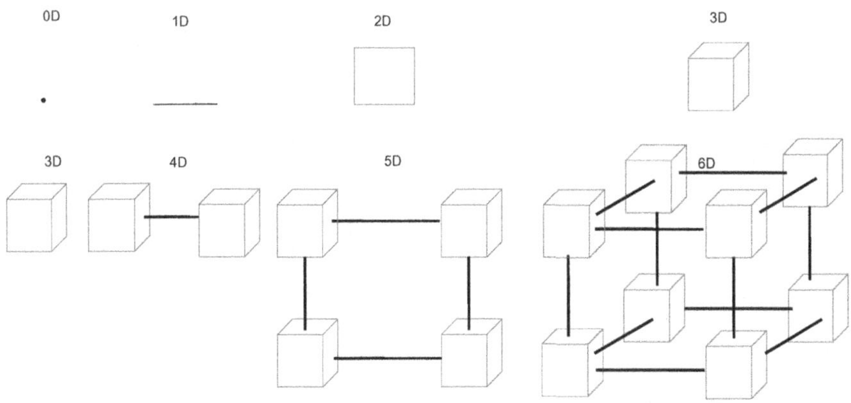

Fig. 8.2 General Periodic Table

Here objects as o1, o2 are separated on the basis of properties as a, b, c in installations as A, B and C corresponding to different conditions. It is an FCA application and generalization to higher dimensionality.

Figure 8.6 shows the polytope 6D for classification

Figure 8.7 shows Polytopic Roadmaps for multi-scale transfer equipment.

Figure 8.8 shows the Periodic Table for multi-scales transfer equipment.

It is the reorganization and development of the roadmaps shown in Fig. 8.7. The columns correspond to lines, squares and cubes

Figure 8.9 shows Polytopic Roadmaps for industry 4.0 and beyond.

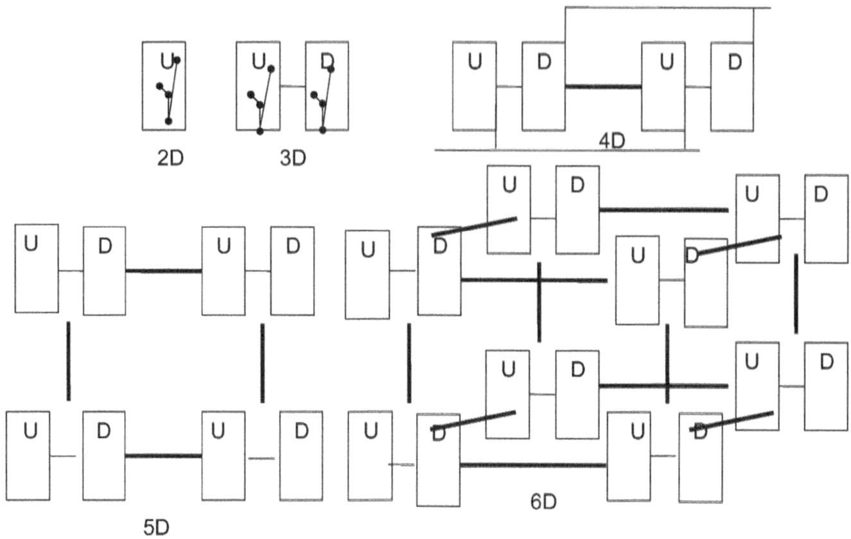

Fig. 8.3 Polytopic Roadmaps for dual graded graphs

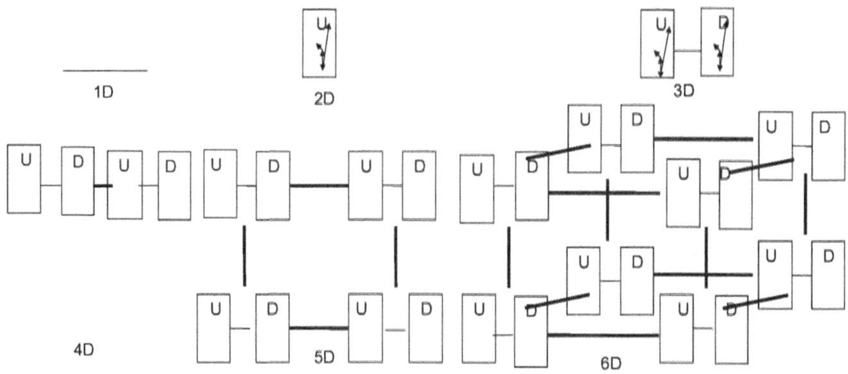

Fig. 8.4 Periodic Table for dual graded graphs

It is based on industry 4.0 toolbox.
The basic levels in Fig. 8.9 are denoted: Integration-Int, Communication-Com, Monitoring-Mon, and IT Service-IT.
Figure 8.10 shows the polytope 7D for industry
We selected a conservative strategy for development.
Int', Com', Mon', and IT' denote the modified basic levels.
Figure 8.11 shows Polytopic Roadmaps for society.
It is a 6D polytope
The basic levels in Fig. 8.11 are Nature, Learning, Labor and Economy.

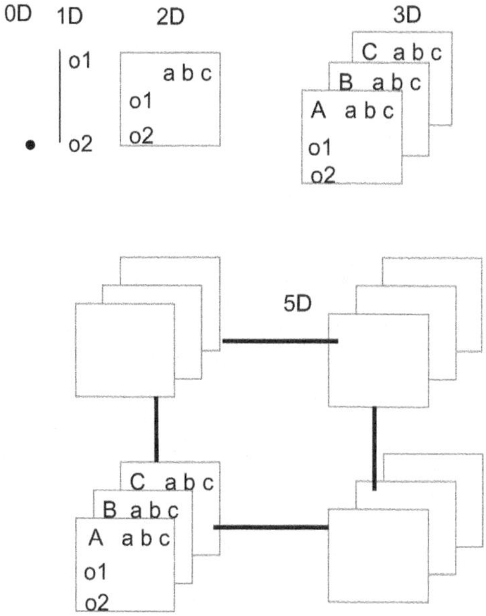

Fig. 8.5 Polytopic Roadmaps for classification: 5D

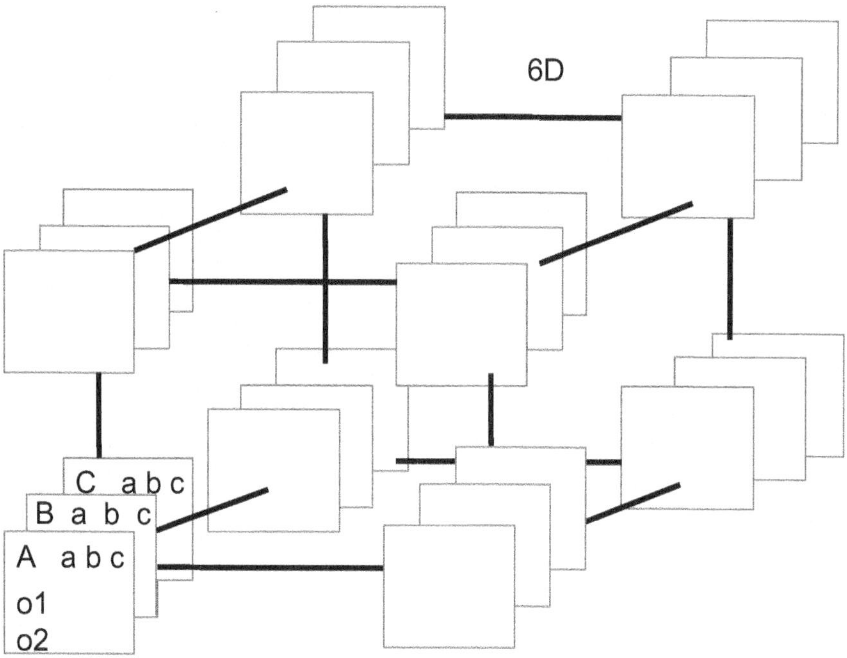

Fig. 8.6 Polytope 6D for classification

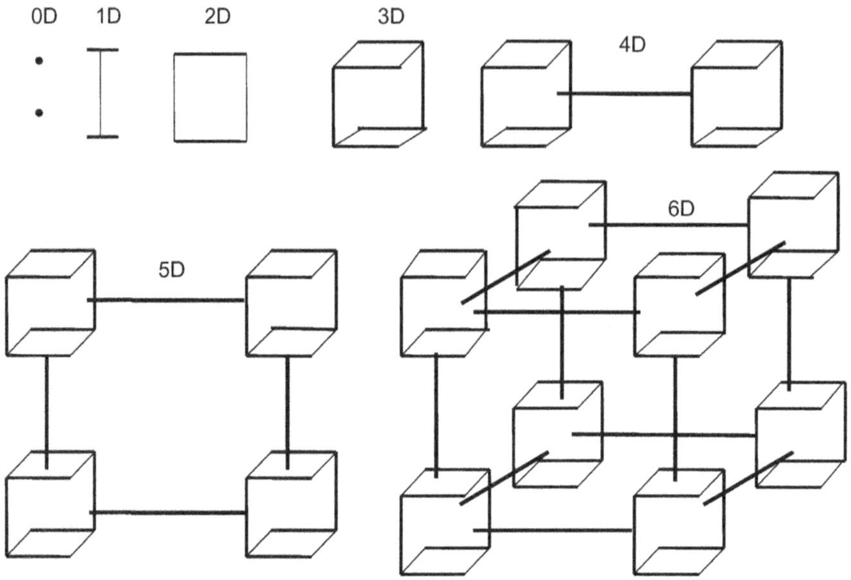

Fig. 8.7 Polytopic Roadmaps for multi-scale transfer: 6D

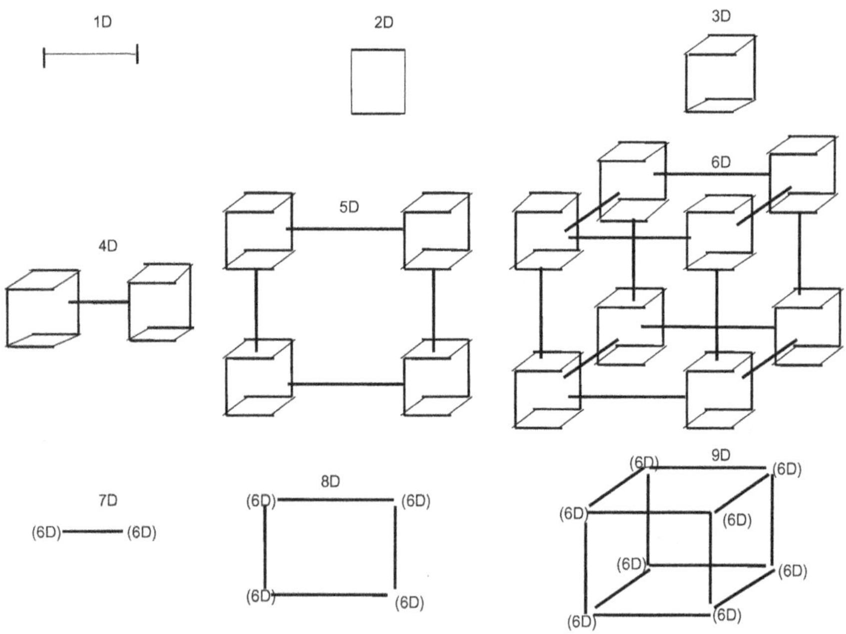

Fig. 8.8 Periodic Table for multi-scales transfer equipment

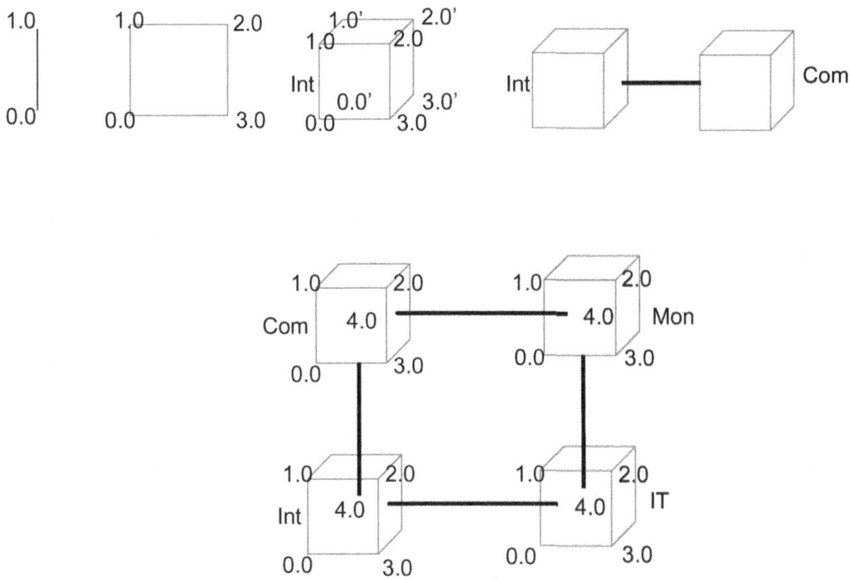

Fig. 8.9 Polytopic Roadmaps for industry: 6D

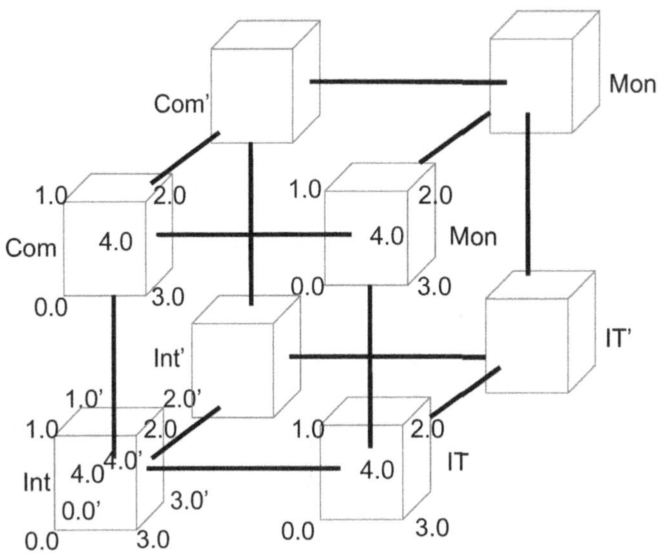

Fig. 8.10 Polytope 7D for industry

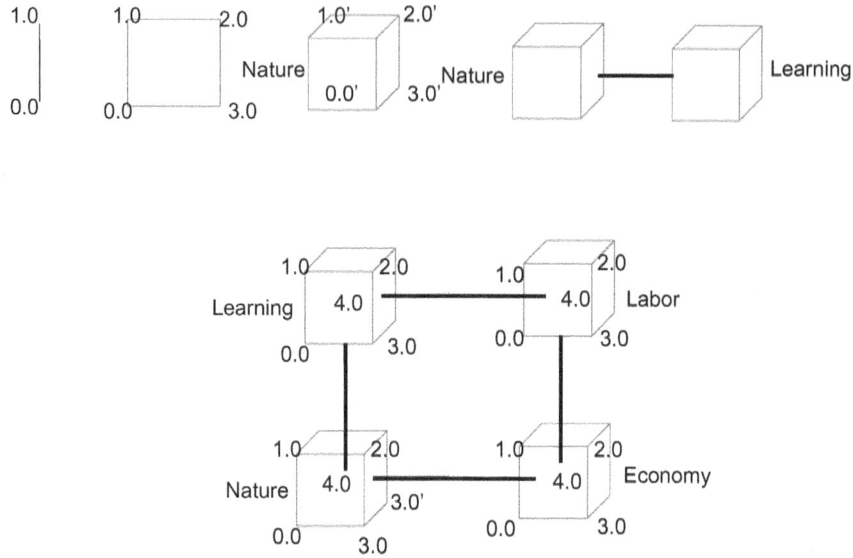

Fig. 8.11 Polytopic Roadmaps for society: 6D

We selected a conservative strategy for development.

The modified basic levels in Fig. 8.12 are Nature', Learning', Labor' and Economy'.

The 7D framework is presented in Fig. 8.12.

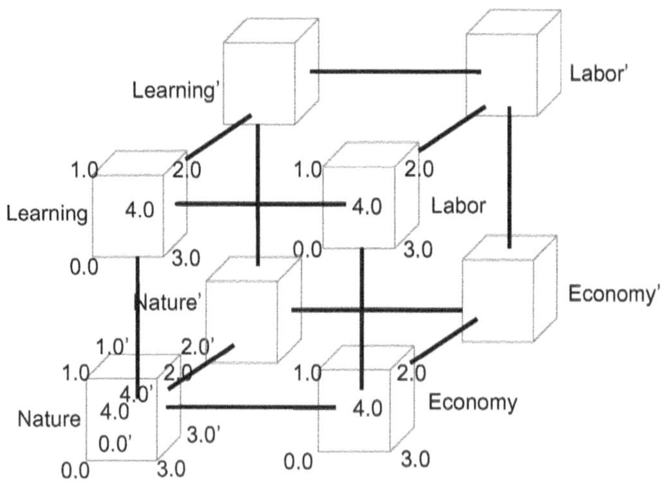

Fig. 8.12 Polytope 7D for society

The Roadmaps shown in Fig. 8.1, 8.2, 8.3, 8.4, 8.5, 8.6, 8.7, 8.8, 8.9, 8.10, 8.11 and Fig. 8.12 outline a high degree of similarity.

This suggests that the doubling and relating may be considered as a basic mechanisms for building universal roadmaps proposed as guidelines, as common methodologies for engineers, researchers and entrepreneurs from various specialties to synthesize their thinking and capabilities into new projects implementation.

The source of inspiration for this similarity is the general developments stages theory as presented by neo-Piagetian theories [9, 12].

This concerns the engineering project of extending the capacity of operations, equipments, technologies or industries to perform functions that would be associated to cognitive-intelligent tasks if performed by people.

The basic endeavor is to outline and implement the isomorphism between the cognitive-intelligent development stages, and the smart operations, equipments, technology and industry development stages. It is this general isomorphism with respect to the higher dimension development and cyclic control mechanisms that explain the similarity of the studied roadmaps. The framework generality illustrate the reality inspired engineering, RIE capability.

8.2 Polytopic Projects

8.2.1 Material Systems

According to Hartmann [13] there exist four basic levels of reality: material, biological, cognitive and intelligent levels.

Leading technological and scientific projects targeting, new materials, artificial living artificial cognitive and artificial-intelligent systems will be presented in what follows.

The presentation aims to highlight the projects basic similarity despite the noticeable difference as domains of application.

The project of polytopic chemistry is considered as promising for new materials.

Figure 8.13 shows a roadmap for Metal Organic Framework (MOF) development [32].

Here the square complexes are considered as intermediary stage [10, 11].

Figure 8.14 shows a polytope 4D associated to MOF developments.

The notations from Fig. 8.14 are:

- S-Molecules
- K1-Polymers
- K2-Squares
- K3- MOF.

The traditional objects of chemical studies were the molecules, denoted here by dots and the polymers considered as 1D.

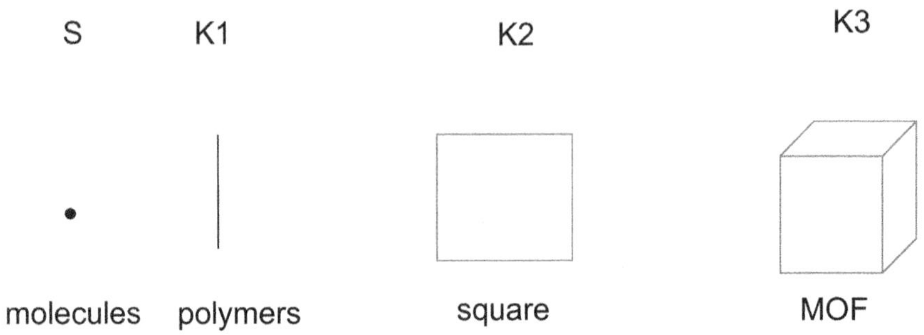

Fig. 8.13 Polytopic Roadmap for MOF development

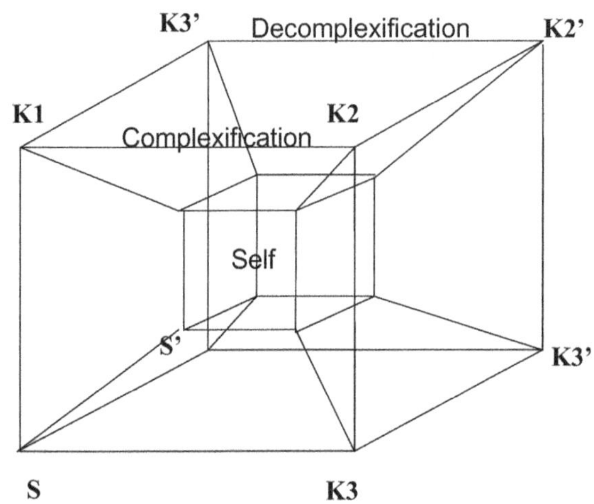

Fig. 8.14 Polytope 4D for MOF development

In the next step, the 2D square complexes have been studied.

A conservative strategy for MOF development shown in Fig. 8.14 is selected.

Complexification and decomplexification may coexist in duality.

During the dual processes of complexification decomplexification the diversity allowing self-evolvability may be initiated.

The direct sequence S → K1 → K2 → K3 corresponds to conditioning. The reverse sequence K3′ → K2′ → K1′ → S′ corresponds to deconditioning.

An exploratory strategy would imply reconditioning for reverse sequence that will be: K4 → K5 → K6 → K7.

The development of Polytopic Roadmaps should continue towards nD as is shown in Fig. 8.15. This is a Periodic Table for MOFs.

The 2D squares are included here. The 4D may be that illustrated by cube inside cube or the coupling of two cubic frames.

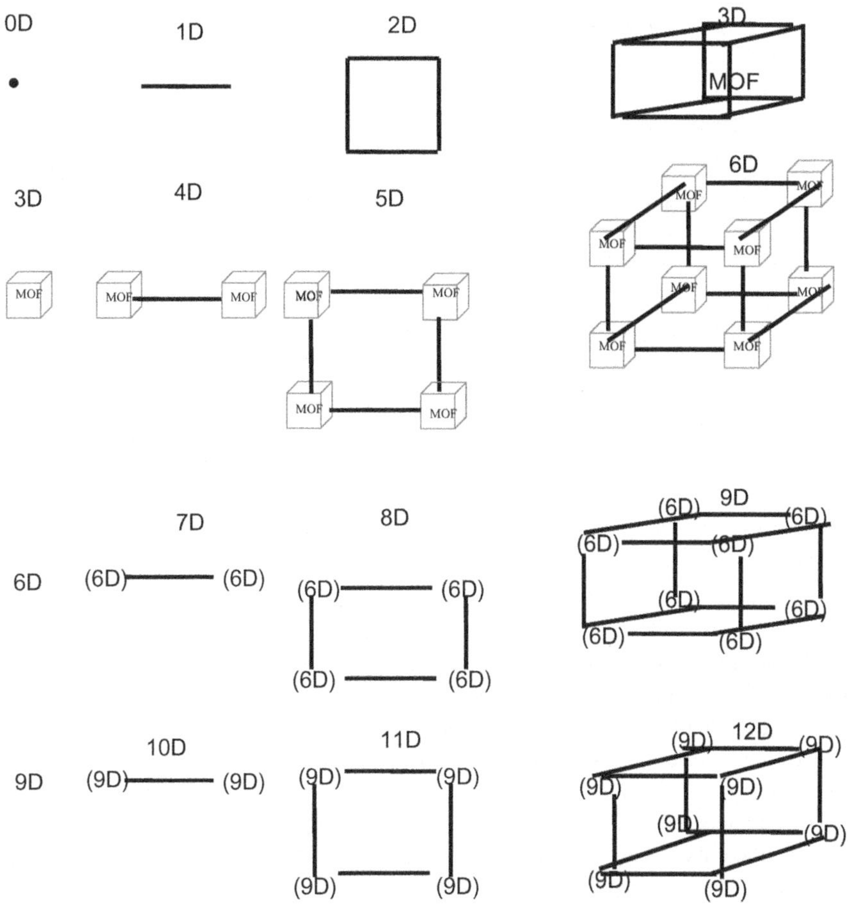

Fig. 8.15 Periodic Table for MOFs-variant 2

The columns correspond in succession to the molecule (dots), polymer (lines), squares or rectangle complexes (squares), MOF (cubes) . It is assumed that the polymers associates in squares and these in turn associate in cubes.

The next row unincluded in this figure will show 12D, 13D, 14D and 15D frames.

Polytopic compounds with cube inside cube (in chemical terms MOF inside MOF) for 4D and higher dimensions are of interest too.

The 4D framework is obtained by joining all the corners of inner cube (MOF) with the corresponding corners of the outer cube (MOF).

Figure 8.16 shows a polytope 4D from the point of view of polytopic chemistry.

It is a line of MOFs as illustrated by Fig. 8.16a. Figures 8.16b, c show detailed presentations. In all cases there are 8 ligands between the two MOF molecules.

Figure 8.16a corresponds to lower density materials while Fig. 8.16c corresponds to higher densities.

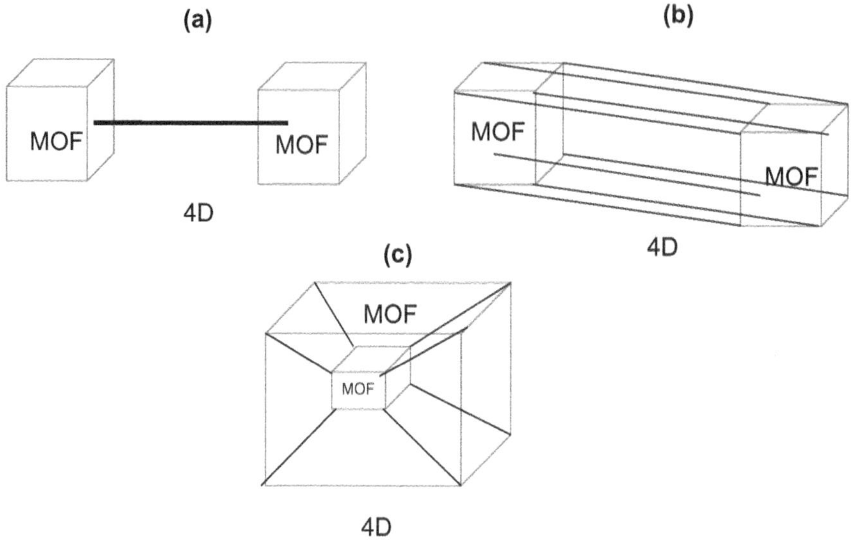

Fig. 8.16 Polytopes 4D for polytopic chemistry

Higher dimensional polytopes are presented in Figs. 8.17 and 8.18.

Figure 8.17 is based on the 4-cube.

Cubelets fill all the corners of the inner and outer cube.

The cubelets are supposed to receive information, analogical or digital and transfer this.

The cubelets may be MOF or other chemical structures.

Since the cubelets are 3D the resulting polytope will be 7D.

It is a 4D-hypercube of 3D-cubes. The dimension is $4 + 3 = 7$.

Fig. 8.17 Polytope based on 4-cube

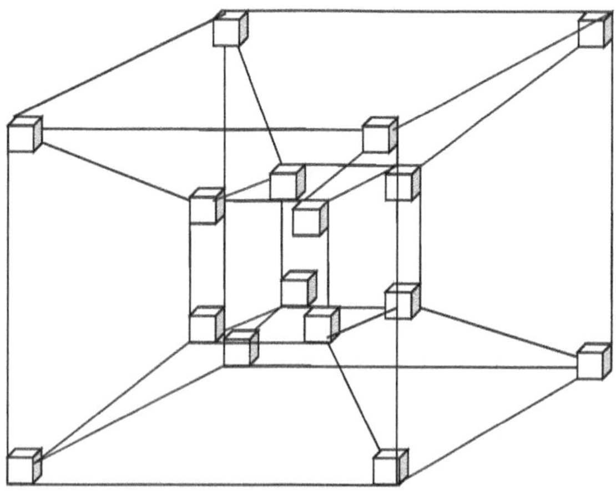

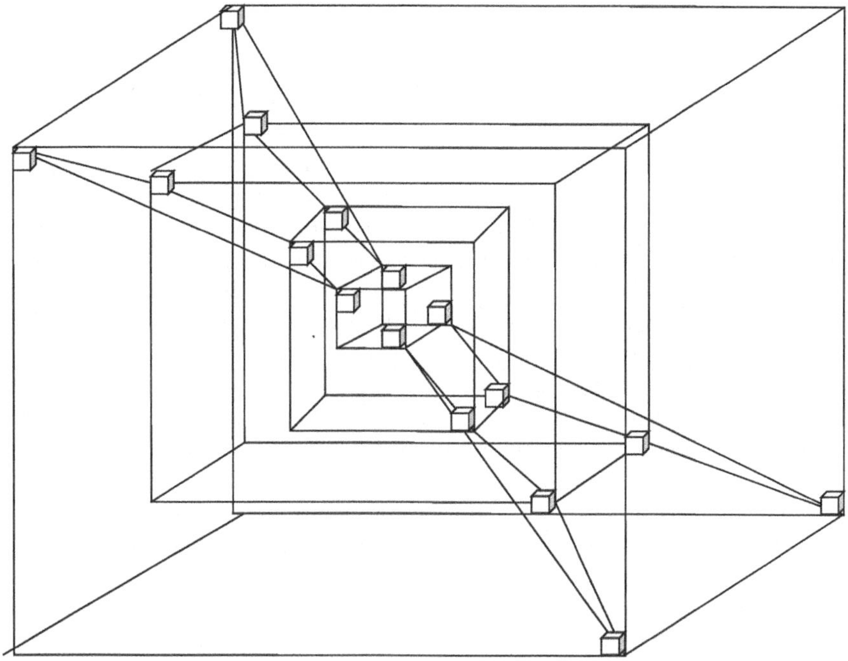

Fig. 8.18 Polytope based on 5-cube

The polytope shown in Fig. 8.18 is based on a 5-cube [18].

For clarity reasons only a part of the 5-cube is represented and decorated with cubelets. The 5-cube potentialities for investigation are considerably increased if compared to 4-cube frameworks. A hierarchy of stability domains, and "Self" modules should be considered.

Since the cubelets are 3D the resulting polytope will be 8D.

It is a 5D-hypercube of 3D-cubes. The dimension is $5 + 3 = 8$.

The challenge is to build molecular structures that are controlled by design. If we can design the structures, we will also be able to tailor the physical properties to match specific needs. It is expected that well-tailored polytopic chemicals with desired properties will be rationally designed and explored for their practical applications.

The fabrication of physical objects is challenged by emerging materials science. Chemists design and build from the molecular level, optimizing features and creating new materials, radically improving quality and reducing waste.

Devices and products are already appearing based on computationally engineered materials that did not existed a few years ago.

New materials will be economically significant when combined with 3D and 4D printing, also known as direct-digital manufacturing-literally printing parts and devices using computational power, and lasers.

Emerging are printed parts for high-value applications like virus specific masks, patient-specific implants or lighter and stronger aircraft parts.

8.2.2 Biological Systems

The dynamical hierarchies of biological systems is investigated here as a promising way towards artificial living systems [6].

Complex, robust functionalities can be generated naturally in at least two ways: by the assembly of structures and by the evolution of structures. Rasmussen et al. [26] studied spontaneous formation of structures. This work defines the notion of dynamical hierarchies in natural systems and shows the importance of this kind of organization for living systems. A framework that enables to formulate, investigate, and manipulate such dynamical hierarchies was defined. This framework allows investigating different levels of description together with their interrelationship, which is necessary to understand the nature of dynamical hierarchies. The framework was then applied to a concrete and simple formal, physicochemical, dynamical hierarchy involving water and monomers at level one, polymers at level two, and micelles that are polymer aggregates at level three.

Formulating this system as a two-dimensional molecular dynamics lattice allows within one dynamical system to demonstrate the successive emergence of two new higher levels of robust structures with associated properties.

The framework for dynamical hierarchies can be used for predictive physicochemical simulation of molecular self-assembly and self-organization processes.

Rasmussen et al. (2008) discussed the detailed process of micelles formation in 3D and formulated an *Ansatz* for how to generate robust, higher-order emergent properties in formal dynamical systems that is based on the conjecture of a necessary minimal complexity within the fundamental interacting structures once a particular simulation framework is chosen.

Figure 8.19 shows a roadmap for dynamical hierarchies

Figure 8.20 shows the polytope 4D associated to dynamical hierarchies. It is the basic cognitive frame for self-evolvability in which the states S interact in a specific way with K1, K2, and K3.

The notations from Fig. 8.20 are:

- S-Molecules

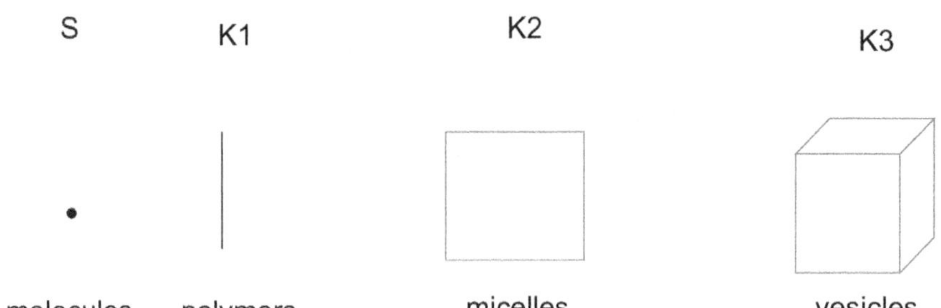

Fig. 8.19 Polytopic Roadmap for dynamical hierarchies

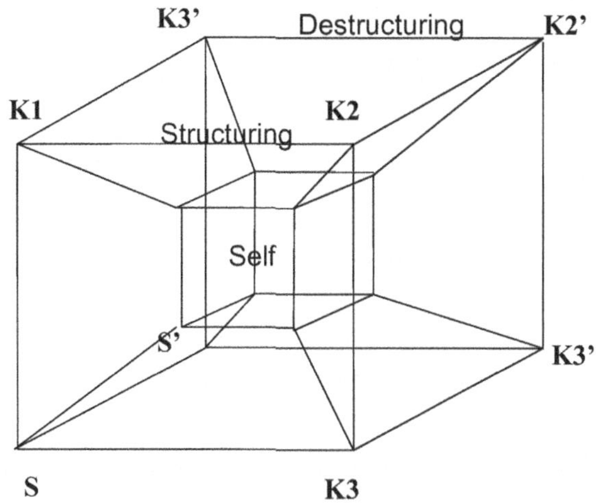

Fig. 8.20 Polytope 4D for dynamical hierarchies

- K1-Polymers
- K2-Micelles
- K3- Vesicles.

The traditional objects of study were the states, S denoted here by dots. Then the interest shifted from states to 1D polymer.

In the next step of modeling the conditioned processes have been introduced. These imply the study of micelles.

Studies of self-evolvable systems show that if there is a process there is a counter-process too. The same is valid for meta-processes and for meta-meta processes. Structuring and destructuring, conditioning and deconditioning, assembly and disassembly, composition and decomposition, construction and deconstruction should coexist in duality.

We selected a conservative strategy for dynamical hierarchies.

Implementing polytopic projects may start from the direct sequence of structuring $S \rightarrow K1 \rightarrow K2 \rightarrow K3$ and complements this by the reverse sequence of destructuring $K3' \rightarrow K2' \rightarrow K1' \rightarrow S'$. This reverses the arrows direction in K1, K2 and K3. Variation of the direct sequence during reverse processing may happen.

Conditioning from the direct sequence is replaced by deconditioning in the reverse sequence. It is during these dual processes of conditioning and deconditioning that the self-evolvability conditions may be generated.

The development of Polytopic Roadmaps should continue towards nD as shown in Fig. 8.21 that contains a Periodic Table for dynamical hierarchies.

Columns correspond to dots, lines, squares and cubes as basic patterns or skeletons.

For dynamical hierarchies, the first row is referring to molecules, polymers, micelles and vesicles denoted also by V.

Second rows start from vesicles V and organize them in 4D, 5D and 6D frames.

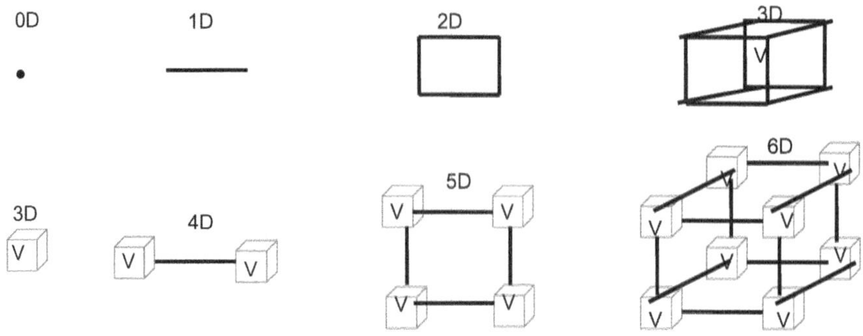

Fig. 8.21 Periodic Table for dynamical hierarchies

The 4D, 5D and 6D represent interacting vesicles and may be considered also as vesicles insides other vesicles.

Self-organization of polymers and micelles reaches a second level in the self-organization of objects that are themselves self-organized. Vesicles are of special interest in this respect, because compartmentalization must have played a major role in the self-organization of complex matter and thus in the evolution of living cells and organisms.

One may envisage the controlled build-up of architecturally organized and functionally integrated multi-vesicular systems toward the design of artificial cells and polymers and micelles systems of tissue-like character, implementing specific intravesicular and intervesicular processes. The manipulation of the features of vesicles and their behavior is a step in this direction. Thus, liposomes decorated with recognition groups, recosomes, present features such as selective interaction with molecular films, aggregation, and fusion [20].

Dynamical hierarchies represent a potential road towards living technologies.

Living technologies are promising because they share the fundamental properties of living systems [2, 6]. These include self-assembly, self-organization, growth and division, purposeful actions, adaptive complexity, evolution, and intelligence. Existing technologies are becoming increasingly like-alive, and powerful.

Examples of living technology projects are synthetic biology attempts to make living systems from scratch in the laboratory, systems exhibiting collective and swarm intelligence distributed across the world wide web, self-reconfiguring robots, and others.

8.2.3 Cognitive Systems

The neural coding problem for the cognitive systems is considered here as a basic step towards artificial brain.

Neuroscientists try to decipher the neural codes of the brain by searching for reliable correlation between firing patterns of neurons and behavioral functions [21].

As early as the 1920s, Edgar Adrian in his pioneering recording showed that the firing rate of a muscle stretch receptor increases as a function of the weights on the muscle suggesting that information is conveyed by specific firing patterns of neurons.

Two leading neural coding theories can be found in the literature: a rate code and a temporal code [7].

In the rate code, all the information is conveyed in the changes of the firing of the cell.

In the temporal code, information is also conveyed in the precise inter-spike intervals.

Due to the large response-variability at the single-neuron level in the brain, even in response to identical stimuli, these two single-neuron-based decoding schemes often produce significant errors in predictions about the stimulus identities or external information.

Table 8.1 shows the spike trains and the associated codewords [3].

It shows the spike trains corresponding to four simultaneously recorded neurons.

Neurons correspond to rows in this table. One can infer subsets of neurons that co-fire.

Data as shown in Table 8.1 may be studied as formal context tables in FCA [8].

If these neurons are place cells then the first codeword 1110 indicates that the neurons 1, 2, 3 co-fire while the third codeword 0101 tells that neurons 2 and 3 co-fire.

The objects are the four neurons corresponding to rows and the attributes are eight time intervals corresponding to columns.

Figure 8.22 shows an example of neural code roadmap.

It is based on codewords as shown in Table 8.1.

Table 8.1 Spike-trains and codewords

1	0	0	1	1	0	1	0
1	0	1	1	1	1	1	0
1	0	0	0	1	0	0	1
0	1	1	1	0	1	0	0

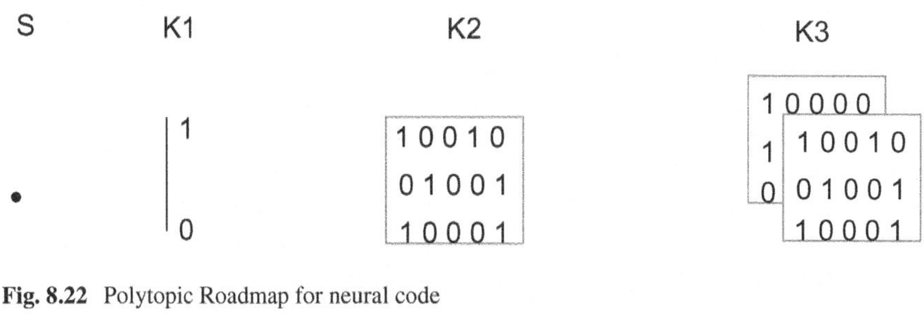

Fig. 8.22 Polytopic Roadmap for neural code

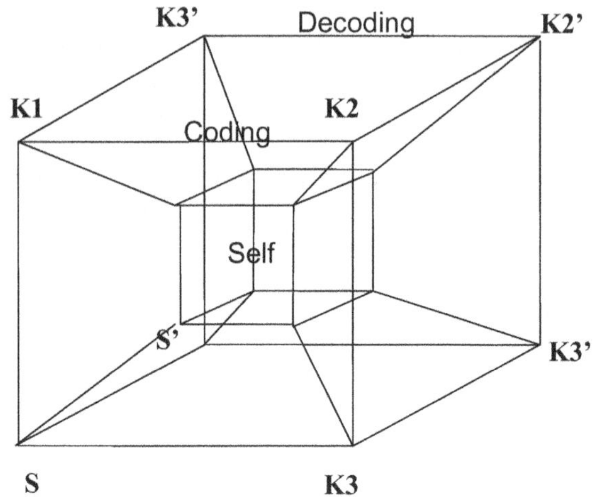

Fig. 8.23 Polytope 4D for neural coding

K2 corresponds to a codewords table, while K3 corresponds to superposition of codewords in a codewords cube.

Figure 8.23 shows a polytope 4D for neural coding.

A conservative strategy is selected for neural coding.

The direct sequence S → K1 → K2 → K3 corresponds to coding. The reverse sequence K3′ → K2′ → K1′ → S′ corresponds to decoding.

Coding from the direct sequence is replaced by decoding in the reverse sequence. It is during these dual processes of coding and decoding that the conditions allowing self-evolvability may be generated.

A possible exploratory strategy would be based on coding and recoding that modify radically the reverse sequence.

Figure 8.24 shows a Periodic Table for neural code.

The columns correspond in succession to: one time interval (lines), neuron-time tables (rectangles), and several tables of neuron-time (cubes) that is sets of parallel contexts.

The first row starts from one line and finish with a 3D-cube.

The second row starts from line relating two cubes (4D) and finish with cube of cubes represented as 6D-cubes.

The third row starts from line relating two 6D-cubes (7D) and finish with 9D-cubes (cube of cubes of cubes).

Column or rows in the context tables may interact according to the rules described by the tick-lines.

The high-dimensionality implications for neural code and artificial brain have been examined by Reimann et al. [29]. Outside of physics, higher dimension is considered just a way of describing complexity. A group of four neurons has dimension three, five neurons dimension four and so on. The more neurons in a group, the higher are the dimension-and so the system becomes increasingly complex. In such studies, dimension does not describe spatial dimensions, but rather the topological dimension

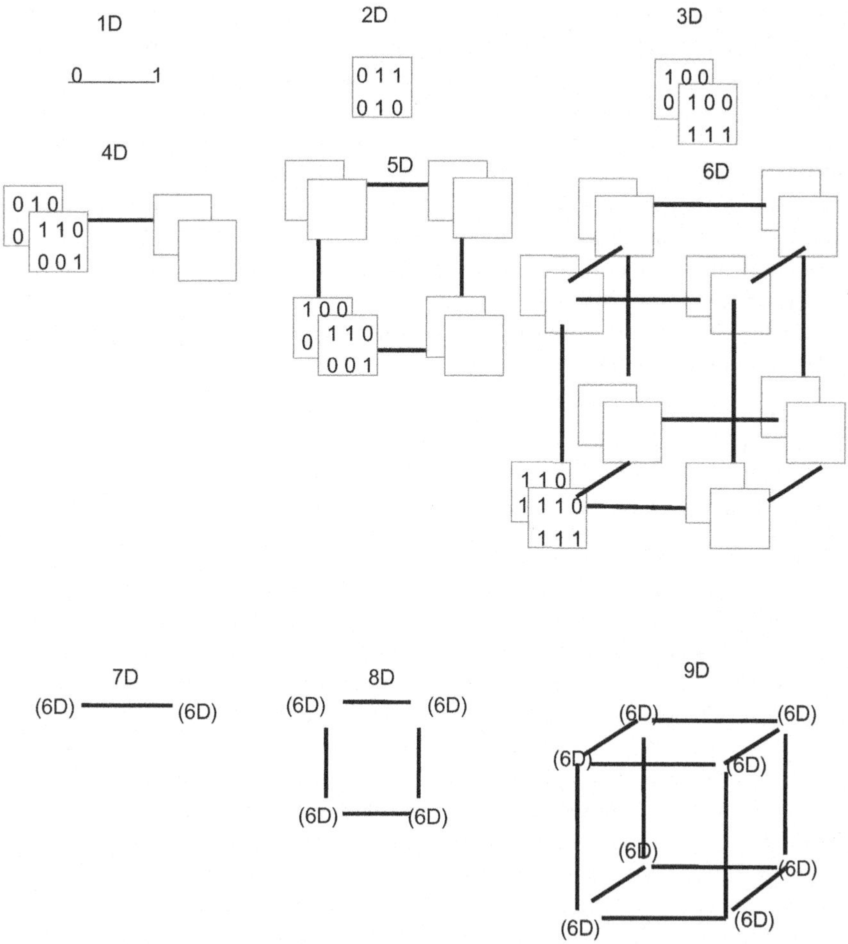

Fig. 8.24 Periodic Table for neural code

of the geometric objects we are describing. The observed 7D simplexes or 11D simplexes are still embedded in the physical 3D space.

8.2.4 *Intelligent Systems*

8.2.4.1 Modeling and Simulation

Modeling pertains to intelligence level. The digital twin (DT) model is a significant project for modeling and simulation complex systems.

Since the reality is structured in levels, associated to multiple scales it is expected that the modeling methods will adopt a similar hierarchical or network architecture.

Four level structures were proposed by the object management group, OMG, to describe informational systems. OMG is an organisation for the standardization in the object oriented field of the software development in which many of the leading software production companies participate [22], OMG 2008 [25]).

For OMG applications in statistics the 0-level refers to data, the 1-level to Models that is models of data, the 2-level to methodologies that is to Meta Models and the 3-level to methodologies that define methodologies that is to Meta Meta Models.

Figure 8.25 shows a roadmap for simulation and modeling.

This describes the development stages for modeling and simulation.

The notations from Fig. 8.25 are:

S-Data
M1-Models
M2-Meta Models
M3- Meta Meta Models.

S denoted also M0 corresponds to data.

M1 models correspond to individual applications. Simulation is limited to specific topics by expert for instance mechanics, mixing, kinetics and so on. Experts use such models to understand and predict phenomena, to optimize or to perform failure analysis. This stage is represented in Fig. 8.25 as a 1D segment suggesting that M1 is able to select the data represented as a dot and predict the road to be followed.

M2 Meta Models correspond to simulation tools. Simulation is a standard tool to answer specific design and engineering for instance, fluid mechanics, thermodynamical analysis, flow-sheeting and so on. The 2D representation suggests that M2 is able to run several M1 models and to connect them on a page. Computer aided design, (CAD) and computer aid in product design and engineering may be classified here. The simulation tools allow design validation, optimization and decision support.

M3 Meta Meta Models correspond to simulation-based system design. Simulation allows a systemic approach to multi-level and multi-disciplinary systems with enhanced range of applications as for instance model based systems engineering

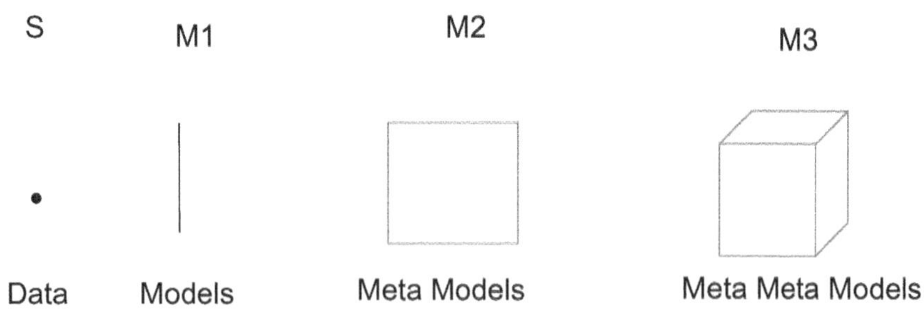

Fig. 8.25 Polytopic Roadmap for modeling

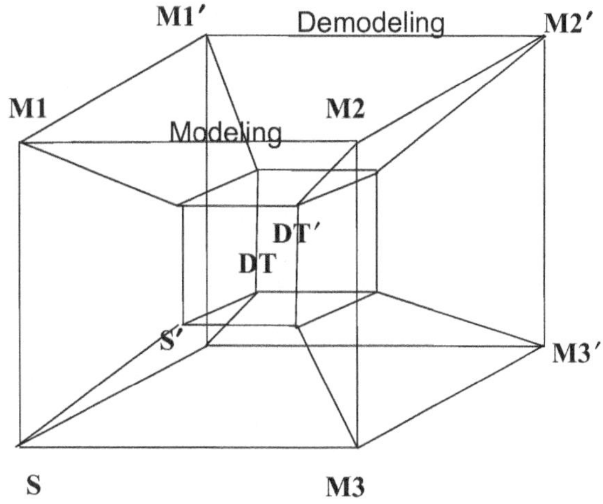

Fig. 8.26 Polytope 4D for modeling

(MBSE). They are critical for communication across departments and companies. Product life cycle management (PLM) may be classified here. The 3D representation suggests that M3 is able to put together several M2 pages of models and results grouping them in a cube.

This level is surpassed now by digital twins (DT). These increase dimensionality to 4D and close the loop.

Figure 8.26 shows a polytope 4D for modeling.

The DT simulation is a core functionality of systems by means of complete assistance along the entire life cycle. An example is supporting operation and service with direct linkage to operation data. Here we need to include business driver bridging different value chains, generic designs and models closing the loop. DT means that M3 close the loop to data S [24].

Implementing DT may start from the direct sequence of modeling S → M1 → M2 → M3 and evaluate this by the reverse sequence of demodeling: M3′ → M2′ → M1′ → S′.

DT' controls the demodeling.

We selected a conservative strategy based on modifications of the direct sequence during reverse processing.

The direct sequence S → K1 → K2 → K3 corresponds to modeling. The reverse sequence K3′ → K2′ → K1′ → S′ corresponds to demodeling.

Demodeling allows focusing on particular domains.

A selection of an exploratory strategy would replace demodeling by remodeling based on radical changes.

Figure 8.27 outlines a Periodic Table for modeling.

Columns correspond to dots, lines, squares and cubes as basic patterns or skeletons.

For modeling, the first row is referring to data, M1, M2 and M3.

Second rows start from M3 and organize models as 4D, 5D and 6D frames.

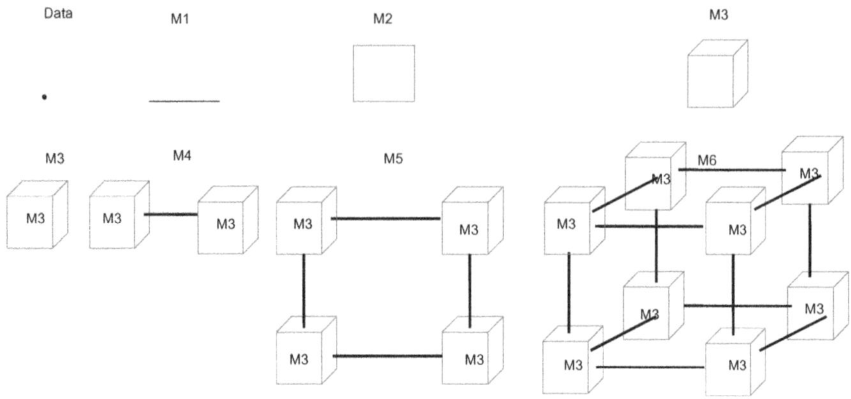

Fig. 8.27 Periodic Table for modeling

M4 corresponds to the digital twin opening.

The 4D, 5D and 6D represent interacting models but may be considered also as models insides other models or models of models.

These geometries have been extensively studied as parallel computer frameworks [1, 30].

8.2.4.2 Generic Models

To elaborate DT we need generic designs and generic models closing the loop between models and data.

The generic models strategy is to maintain the pattern of the conventional real models but to adapt the definitions of parameters, functions and operations to specific features of the studied phenomena.

In this way we have structurally stable models admitting some formulations invariant under the choice of the algebraic structure of numbers for space time and other parameters. The models and methods for complex systems appear as twins of the conventional real field models and methods. The methodology based on twin models, may be considered as a source for artificiality and artifacts.

It may be related to engineering as the artificial chemistry is related to chemistry [5] or artificial life to natural life [2].

In the Newtonian reductionism sense, time and space are independent of each other and of the existing entities and events occurring. Space and time are set in advance and are invariant.

On the contrary, in the Leibnizian systemic sense, the space means order of coexistence, and time means order of succession. Leibnizian space and time are defining by referring to each other, as complementary and dependent on the included entities and events, that is, on specific context.

The point of view adopted for complex systems modeling is that the algebraic frame for variables as time and space must agree first of all with the context of analysis of the studied system. Replacing universal concepts for time and space, by relational concepts as, multi-dimensional, multi-scaled, dyadic, and cyclic time and space are permitted and implemented if the system analysis can proceed on this basis [16]. The reformulation of the models with multi-scaled or cyclic parameter is based on experimental observation of some systems and may be used to redesign other systems in order to become multi-scaled, cyclic and so on [15, 17].

The high-dimensionality and the cyclicity requiring multi-dimensional, multi-scaled and cyclic time are key strategies to face and prevails complexity.

The physical domains offer inspiration for the artificial domains, both for calculus and for artefacts. If we study for instance the mixing in a chemical reactor by detailed models it is possibly to be surpassed by calculus complexity and results interpretation. But introducing a cyclic time and space based on experimental observations, a simplified model may result and offer the basic information. Moreover, we can impose circularity for instance by closing the tracer particles in the reactor. The study of the cyclic residence time distribution for such tracers allows characterizing the mixing in that equipment.

An application of generic models methodology is considered more in detail now.

It refers to the first order wave equation representing the material balance for transport processes.

The real field wave model shows that the variation of concentration, in time and space, is due to convective processes and to kinetic processes of interaction.

The model (8.1) describes phenomena of major physical and technological interest, as momentum, heat and mass transfer accompanied by chemical reaction.

$$\frac{\partial y}{\partial t} + v\frac{\partial y}{\partial z} + q(y) = 0 \tag{8.1}$$

Here y (t, z) may denote an concentration variable in t-time and z-space, v denotes the velocity, and q(y) denotes the interaction rate [28].

All the parameters of the model (8.1) are defined on the mathematical real field.

For physical systems study, the real field plays the dominant role.

However this field may be unsuitable to control in real time some aspects of high complexity systems.

The parameters and the functions of the real field conventional differential models for transfer have to be replaced by different algebraic structures to take into account the specificity of that transfer.

The Galois field GF (m) where m is the number of elements of the field represents mathematical frameworks that have been used to describe the finite, logical or cyclic type of processes. If m is not a prime number we are faced with rings instead of fields.

An example, the sum and product operations in GF (4) are presented in Table 8.2

The wave equation, WE, equation useful for complex systems study will be described in what follows [15].

Table 8.2 Sum and product in GF (4)

(x⊕y)					(x⊗y)				
⊕	00	01	10	11	⊗	00	01	10	11
00	00	01	10	11	00	00	00	00	00
01	01	00	11	10	01	00	01	10	11
10	10	11	00	01	10	00	10	11	01
11	11	10	01	00	11	00	11	01	10

We use the space $Z = z_0z_1 \ldots . z_j$ interpreted as features space, the time $T = t_0t_1 \ldots t_j$ interpreted as the degree of pattern recognition, classification or development, the output $Y = y_0y_1y_2 \ldots y_j$, the interaction rate $Q = q_0q_1q_2 \ldots q_j$ and the velocity $V = v_0v_1v_2 \ldots v_j$.

Here z_j, t_j, y_j, q_j and v_j are digits, 0 or 1.

One of the simplest mechanisms of pattern recognition, classification or development is that in which small changes of the degree of pattern recognition, ∂T, are associated to small changes of the output, ∂Y. It should be emphasized that the differential is in fact a difference since T and Y are discrete.

Moreover, the change of output ∂Y depends on both the existing output Y and the change ∂T of T that is:

$$\frac{\partial Y}{\partial T} \propto Q(\mathrm{Y}) \tag{8.2}$$

It is supposed that ∂T is non-null. Otherwise the differential equations will be replaced by difference equations.

Another pattern recognition, classification or development mechanism, takes into account that the variation of the output Y, along the degree of pattern recognition T is proportional to the output variation along the features space Z. Classification, pattern recognition and development means in fact travel in time T, along the space of properties, Z. As Z is screened with a velocity V, the output varies proportionally.

This means that:

$$\frac{\partial Y}{\partial T} \propto V \otimes \frac{\partial Y}{\partial Z} \tag{8.3}$$

Here the velocity is a vector V pertaining to the same field as Z, T, Q or Y.

This mechanism is of convection or drift type.

The general model of the pattern recognition or classification process including both types of recognition or classification processes, corresponding to the evolution according to T and Z, is the first order wave equation WE:

$$\frac{\partial Y}{\partial T} \oplus V \otimes \frac{\partial Y}{\partial Z} \oplus Q(Y) = 0 \tag{8.4}$$

The initial condition is:

$$Y(Z, 0) = F(Z) \tag{8.5}$$

Obviously V and Q may depend on T and Z.

The fact that the addition is equivalent to the difference suggests that a second order wave equation doesn't gives new solutions.

The first-order wave equation, WE, is formally similar to the model (8.1) extensively applied in different domains by chemical engineers.

Models (8.1) and (8.4) are twins.

For Q = 0 the wave equation became:

$$\frac{\partial Y}{\partial T} \oplus V \otimes \frac{\partial Y}{\partial Z} = 0 \tag{8.6}$$

It is the so-called convection or plug-flow model.

A solution that can be verified directly with (8.6) is the characteristic:

$$Y = Z \oplus V \otimes T \tag{8.7}$$

Table 8.3 shows the values of V⊗T:

For different velocities V the solutions (8.7) has been shown in Tables 8.4, 8.5 and 8.6.

These tables contain Latin Squares.

Table 8.3 V⊗T

V⊗T				
⊗	00	01	10	11
00	00	00	00	00
01	00	01	10	11
10	00	10	11	01
11	00	11	01	10

Table 8.4 Convection Model, V = 01

Z⊕ V⊗T				
⊕	00	01	10	11
00	00	01	10	11
01	01	00	11	10
10	10	11	00	01
11	11	10	01	00

Table 8.5 Convection Model, V = 10

Z⊕ V⊗T				
⊕	00	10	11	01
00	00	10	11	01
01	01	11	10	00
10	10	00	01	11
11	11	01	00	10

Table 8.6 Convection Model, V = 11

Z⊕ V⊗T				
⊕	00	11	01	10
00	00	11	01	10
01	01	10	00	11
10	10	01	11	00
11	11	00	10	01

Table 8.7 Latin square

0	1	2	3
1	0	3	2
2	3	0	1
3	2	1	0

For example, replacing binary representation by integers, Table 8.4 became (Table 8.7):

Latin squares are distinctive DoE.

For null velocity V the first order wave equation, WE, reduces to the kinetic model:

$$\frac{\partial Y}{\partial T} \oplus Q(Y) = 0 \tag{8.8}$$

The solution is written:

$$Y = \int Q\mathrm{d}T = Q * T \tag{8.9}$$

This integral appears as a particular type of product "*".

Let us define the integral:

$$Y = \left[\sum_j q_j t_j\right] \mathrm{mod}2 \tag{8.10}$$

Table 8.8 Kinetic Model

(Q*T)				
*	00	01	10	11
00	0	0	0	0
01	0	1	0	1
10	0	0	1	1
11	0	1	1	0

Table 8.9 Kinetic Model modified

1	1	1	1
1	−1	1	−1
1	1	−1	−1
1	−1	−1	1

The integral definition with the mod 2 operation is selected to ensure the logical interpretation of the output Y [15].

Table 8.8 shows the results of the kinetic model

Replacing 0 by 1 and 1 by −1 in Table 8.8 we obtain an equivalent solution in the form of Walsh-Hadamard waves. As the Latin squares these are distinctive DoEs (Table 8.9).

We observe that the solutions of the model (8.4) are designs of experiments.

The results interpretation is that the digital twin model (8.4) shows how to control the solutions of the real model (8.1).

The complexity control needs both models.

References

1. Ammon, J.: Hypercube Connectivity within ccNUMA Architectures. SGI Origin Team, Mountain View, CA (1998)
2. Bedau, M.A., McCaskill, J.S., Packard, N.H., Rasmussen, S.: Living technology: Exploiting life's principles in technology. Artif. Life **16**(1), 89–97 (2010)
3. Curto, C.: What can topology tell us about the neural code? Bull. Am. Math. Soc. **54**(1), 63–78 (2017)
4. DeLisi, P.S.: The Glass bead Game Linking Interdependence and organizational learning (1999). https://www.org-synergies.com/docs/Glass-Bead-Game.pdf
5. Dittrich, P., Ziegler, J., Banzhaf, W.: Artificial chemistries—a review. Artif. Life **7**(3), 225–275 (2001)
6. Dorin, A., McCormack, J.: Self-assembling dynamical hierarchies. Artif. Life **8**, 423–428 (2003)
7. Eggermont, J.J.: Is there a neural code? Neurosci. Biobehav. Rev. **22**(2), 355–370 (1998)
8. Endres, D.M., Földiák, P., Priss, U.: An application of formal concept analysis to semantic neural decoding. Ann. Math. Artif. Intell. **57**(3–4), 233–248 (2009)
9. Fischer, K.W.: Dynamic cycles of cognitive and brain development: measuring growth in mind, brain, and education. The Educated Brain: Essays in Neuroeducation, pp. 127–150 (2008)

10. Fujita, M.: Metal-directed self-assembly of two-and three-dimensional synthetic receptors. Chem. Soc. Rev. **27**(6), 417–425 (1998)
11. Fujita, M., Tominaga, M., Hori, A., Therrien, B.: Coordination assemblies from a Pd (II)-cornered square complex. Acc. Chem. Res. **38**(4), 369–378 (2005)
12. Halford, G.S., Andrews, G.: Information—processing models of cognitive development. The Wiley-Blackwell Handbook of Childhood Cognitive Development, vol. 16, pp. 697–722 (2010)
13. Hartmann, N.: The New Ways of Ontology. Greenwood Press, Westport (1952)
14. Hesse, H.: The glass bead game. Holt, Rinehart and Winston, Inc., New York (1969)
15. Iordache, O.: Theoretical frames for smart structures. Mater. Sci. Eng., C **4**(3), 143–148 (1996)
16. Iordache, O.: Polystochastic Models for Complexity. Springer, Berlin, Heidelberg (2010)
17. Iordache, O., Valentin, G., Corriou, J.P., Pons, M.N., Pethö, A.: Intermittent interfacial transfer. ADyadic Model. Acta Chim. Hung. Models Chem. **1**(130), 1–18 (1993)
18. Joswig, M., Ziegler, G.M.: A neighborly cubical 4-polytope, Electronic Geometry Models, No. 2000.05.003, C45 Master.poly. (2000)
19. Klir, G.J.: Trends in General Systems Theory. Wiley-Interscience, New York (1972)
20. Lehn, J.M.: Towards complex matter: supramolecular chemistry and self-organization. Eur. Rev. **17**(2), 263–280 (2009)
21. Lin, L., Osan, R., Tsien, J.Z.: Organizing principles of real-time memory encoding: neural clique assemblies and universal neural codes. Trends Neurosci. **29**(1), 48–57 (2006)
22. Mellor, S.J.: MDA distilled: principles of model-driven architecture. Addison-Wesley Professional (2004)
23. Miller, J.G.: Living Systems. McGraw-Hill Book Company, New York (1978)
24. Negri, E., Fumagalli, L., Macchi, M.: A review of the roles of digital twin in cps-based production systems. Procedia Manuf. **11**, 939–948 (2017)
25. OMG.: Object Management Group, Software & Systems Process Engineering Meta Model Specification 2.0 (2008)
26. Rasmussen, S., Baas, N.A., Mayer, B., Nillson, M.: Ansatz for dynamical hierarchies. Emergence: Contemporary Readings in Philosophy and Science, p. 305 (2008)
27. Ray, F.: General Systems Theory: A Knowledge Domain in Engineering Systems (2000)
28. Rhee, H.K., Aris, R., Amundson, N.R.: First Order Partial Differential Equations II, Theory and applications of hyperbolic systems of quasilinear equations. Prentice-Hall, Englewood Cliffs, N.J. (1989)
29. Reimann, M.W., Nolte, M., Scolamiero, M., Turner, K., Perin, R., Chindemi, G., Dłotko, P., Levi, R., Hess, K., Markram, H.: Cliques of neurons bound into cavities provide a missing link between structure and function. Front. Comput. Neurosci. **11**, 48 (2017)
30. Seitz, C.L.: The cosmic cube. Commun. ACM **28**(1), 22–33 (1985)
31. von Bertalanffy, L.: General Systems Theory: Foundations, Development, Applications. George Braziller, New York (1968)
32. Yaghi, O.M.: Reticular chemistry in all dimensions. ACS Central Sci. **5**(8), 1295–1300 (2019)

Appendix A
Combinatorial Species

The combinatorialspecies have been introduced by Joyal (Joyal, A.: *Une théorie combinatoire des séries formelles.* Adv Math **42**:1–82 (1981).

Bergeron, F., Leroux, P., Labelle, G.: CombinatorialSpeciesand Tree-Like Structures, Encyclopedia of Mathematics and Its Applications, vol. 67. Cambridge University Press (1998)).

A structure type is just any sort of structure we can put on finite sets: ordering, coloring, permutation, partition, combination, classification, complexification and so on. In combinatorics we count such structures using generating functions. A generating function is a power series where the coefficient of x^n keeps track of how many structures of the given kind we can put on an n-element set. Using these functions, we can figure out the coefficients and get explicit formulas that count the structures in question. The reason this works is that operations on generating functions come from operations on structure types.

Species are classes of labeled structures.

Let B be the category of finite sets with bijections, and E the category of finite sets and functions. A species of structures is a functor from B to E that is, F: B → E.

Given a species F, we obtain:

- For each finite set U a finite set F[U] which is called the set of F-structures on U
- For each bijection σ: U → V a bijection F[σ]: F[U] → F[V], which is called the transport of F-structures along σ.

The functions F[σ] should satisfy the functorial properties:

For all bijections σ: U → V and τ: V → W, F[τ° σ] : F[τ] ° F[σ]
For the identity map Id_U : U → V, F [Id_U] = $Id_{F(U)}$.

Figure A.1 illustrates the transport of F-structure along σ.

The functor F, which produces the structures F[U] and the transport functions F[σ] can be described in different fashions provided that the functorial conditions hold.

O. Iordache, *Polytopic Roadmaps*, Studies in Systems, Decision and Control 368 ,
https://doi.org/10.1007/978-3-030-75630-7

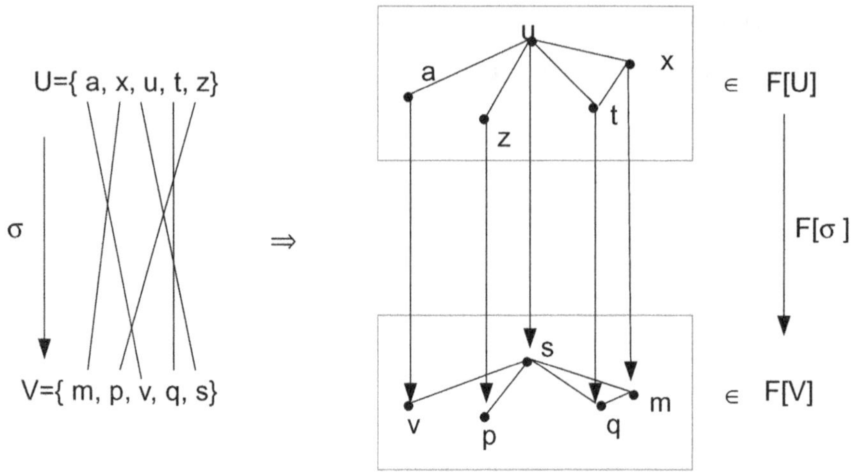

Fig. A.1 Transport of F-structure

For example one can use axiomatic systems, constructions, algorithms, combinatorial operations, functional equations, geometric figures to specify species.

The key to grasp the conceptof combinatorial structure in mathematical terms consists in using the concept of functor, emphasizing the transport of structures along bijections, rather than the properties of the structures.

Appendix B
Formal Concept Analysis

The formalconcept analysis, FCA, is a theory of data analysis which identifies conceptual structures among data sets (Ganter, B., Wille, R.: FormalConcept Analysis. Mathematical Foundations. Springer, Berlin (1999)).

The main goal of FCA has been the support of rational communication and the representation and processing of knowledge based on the so called restructuring program.

A strong feature of FCA is its capability of producing graphical visualizations of the inherent structures among data. The FCA plays a prominent role in conceptual modeling by combining the ease of handling data base object that are defined via a list of properties to a mathematical model rooted in a formalization of logic by which reasoning is based on communicative rationality. In FCA the concept understanding is considered as the basic unit of thought. A particular concept has both an extension and an intension.

FCA supposes that some relation between objects and properties is already established in the form of a context F = (G, M, I) where G is the set of objects, M is the set of properties or attributes and $I \subseteq G \times M$ is the incidence relation between objects and properties.

A formal context F can best be represented by a table specifying which objects fall under which properties. This suggests that a context may be associated to classification purposes.

For a set of objects $A \subseteq G$, we can define all the properties shared by all objects, provided a context F is given:

$$A' = \{m \in M | \forall g \in A : (g,m) \in I\}$$

These are the common properties of A.

Similarly the dual operation can be defined provided a property set $B \subseteq M$ is given:

$$B' = \{g \in G | \forall m \in B : (g, m) \in I\}$$

O. Iordache, *Polytopic Roadmaps*, Studies in Systems, Decision and Control 368 , https://doi.org/10.1007/978-3-030-75630-7

These are the common objects of B.

Assume a context F = (G, M, I) is given. A formalconcept of F is defined as a pair (A, B) where A ⊆ G (called extent) and B ⊆ M (called intent), A′ = B and B′ = A.

Given a context F = (G, M, I) the collection of all formalconcepts$\mathcal{B}$ (G, M, I) forms a complete Galois lattice, GL, where the partial order is defined by (A1, B1) ≤ (A2, B2) iff A1 ⊆ A2 (which is equivalent to B1 ⊇ B2). The supremum and the infimum are defined as follows:

$$\bigvee_{t \in T} (A_t, B_t) = ((\bigcup_{t \in T} A_t)'', \bigcap_{t \in T} B_t)$$

$$\bigwedge_{t \in T} (A_t, B_t) = (\bigcap_{t \in T} A_t), (\bigcup_{t \in T} B_t)'')$$

From an abstract point of view the complete lattice$\mathcal{B}$ (G, M, I) may be interpreted as a category. The operation denoted by:" can be interpreted as a closure operator on both sides the object level, G and the attribute level, M. The closure operation induces a complete lattice and the conceptlattice corresponds to a Galois-connection between two closure operators. A conceptlattice$\mathcal{B}$ (G, M, I) determines a hierarchyof formalconcepts. This hierarchy can be used to perform inferences with respect to properties of concepts or with respect to extensions of concepts.

Back Cover

Polytopic Roadmaps

Octavian Iordache

The paradigm of cognitive-intelligentinspiration introduced and studied here as the basic tool for projects implementation in the domain of complex industrial systems is that of PolytopicRoadmaps.

Polytope is the general term of the sequence point, segment, polygon, polyhedron and so on. The PolytopicRoadmaps are guidelines inspired by the stages of cognitive-intelligentdevelopment, useful for learning, representing, designing, building, operating and controlling of complex engineeringsystems. The objective is to develop engineeringoperations, equipments, technologies and industries in successive stages as humans develop from sensory-motor stage of child to formal and post formaldevelopment stages of adult. The 4D approach developed in previous monographs and correlated to industry 4.0 and fourth industrial revolution is continued here toward higher dimensions.

We identify as polytopicoperations, equipment, technology and industry, the n-dimensional, nD frameworks.

PolytopicRoadmaps are expected to serve as guidelines, to become powerful instruments releasing an abundance of significantly new capabilities and structures.

Dedicated to the practical implementation of projects for high dimensional complex systems, this book is divided in 8 chapters. The book begins with the presentation in Chap. 1 of everything 4.0 frameworks. The higher dimensional, the roadmaps beyond 4.0 are imagined. The role of doublingand relating and of Periodic Tables is outlined. Second chapter contains a summary of methodology that emphasizes on iteration, dimensionality and cyclicality around the center. The cognitive-intelligent source of inspiration, the Piagetian and the neo-Piagetian development stages are briefly presented. Conservative and exploratory strategies for development are introduced.

Chapter 3 presents elementary operations of engineering interest and their relation to combinatorialspecies. Partitions, permutations, classificationsand complexifications are the operations analyzed in this chapter. Latin squares and formalconcept analysis, (FCA), are involved. Chapter 4 is dedicated to equipments able to run the presented high dimensional operations. Multi-scaletransfer equipments, cyclicoperations, conveyorsand assembly lines are the detailed examples. Periodic Tables for equipments are proposed. Chapter 5 presents technologies. Polytopicflow-sheets, on-line analytical processing, (OLAP), polytopicengineeringdesigns, enterprise controlsystemsand realityinspired engineering (RIE) is introduced.

Conceptsas, industry4.0, industry 5.0, and polytopicindustry are presented in Chap. 6. The relation between PolytopicRoadmaps, industry 4.0 toolbox and Periodical Tables is pointed out. Polytopicsociety, society 5.0, represents the subjects of Chap. 7. Bio-Info-Nano-Cogno, (BINC) technologies, Cyber PhysicalSocial Systems (CPSS) and ecology aspects have been discussed in this chapter. Chapter 8 highlights the connection with other systemstheory projects and evaluates the integral perspective towards everything polytopic. The General PolytopicRoadmaps, (GPTR), for growing complexity are proposed as universal guidelines, as common methodologies for engineers, researchers and entrepreneurs from various domains to synthesize their thinking and capabilities into new projects implementation.

Leading polytopicprojects for different levelsof reality are specified.

Introducing ground-breaking fields of most important practical and theoretical interest and roadmaps for the higher dimensional advances in complexity domains, the book will be useful to engineers, researchers, entrepreneurs and students in different branches of production, engineering, systems sciences and applied mathematics.

Keywords Assembly, Classification, Complexification, Conveyors, Cyber PhysicalSystems, Cyclicality, Design of experiments, Developmentstages, DigitalTwin, Engineeringdesign, Equipments, Flow-sheets, FormalConcept Analysis, Higher dimensionality, Industry4.0, Multi-Scale, Operations, Periodic Tables, PolytopicRoadmaps, Society 5.0, Systems science.

Index

O. Iordache, *Polytopic Roadmaps*, Studies in Systems, Decision and Control 368, https://doi.org/10.1007/978-3-030-75630-7

The manufacturer's authorised representative in the EU is Springer Nature Customer Service Centre GmbH, Europaplatz 3, 69115 Heidelberg, Germany. If you have any concerns regarding our products, please contact ProductSafety@springernature.com

Printed and bound by CPI Group (UK) Ltd, Croydon, CR0 4YY
30/07/2026
02182027-0006